Contents

PROGRESSION IN PRIMARY SCIENCE

A Guide to the Nature
and Practice of Science
in Key Stages 1 and 2

Martin Hollins and Virginia Whitby

with Liz Lander, Barbara Parson and Maggie Williams

David Fulton Publishers
London

Published in association with Roehampton Institute London

David Fulton Publishers Ltd
Ormond House, 26–27 Boswell Street, London WC1N 3JD

First published in Great Britain by David Fulton Publishers 1998
Reprinted 1998

British Library Cataloguing in Publication Data
A catalogue record for this book is available from the British Library

ISBN 1–85346–498–8

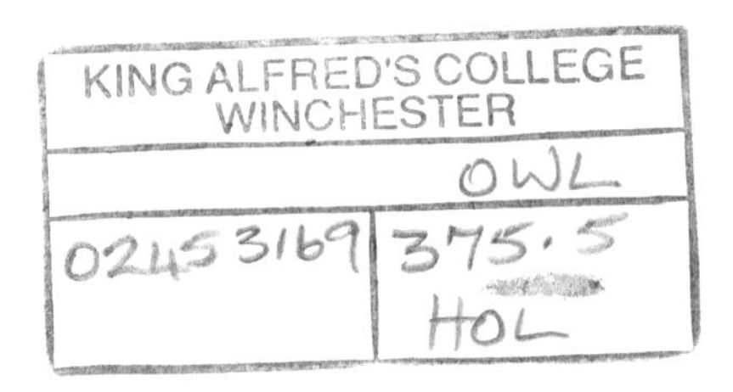

Typeset by FSH Print and Production Ltd
Printed in Great Britain by The Cromwell Press Ltd, Trowbridge, Wilts.

Acknowledgements

We express our thanks to Lynne Bartholomew for her contribution to science in the early years' sections of this book, and to Stephen Cook for his computer expertise and hard labour in producing this book.

This book is dedicated to Philippa, Fiona, Melissa, Lawrence and Robin who may become the scientists of the twenty-first century.

Contributors

Lynne Bartholomew is a senior lecturer in education at Roehampton Institute London and coordinator of Redford House Nursery, situated at Froebel Institute College. She was previously deputy head of a nursery school in Southall, West London, and is co-author with Tina Bruce, of *Getting to Know You: a guide to record-keeping in early childhood education and care.*

Martin Hollins is a senior lecturer in science education at Roehampton Institute London, working in primary and secondary initial teacher education and professional development. Martin is also a freelance education consultant who edits the Association for Science Education's primary journal, *Primary Science Review.* He was formerly the Director of the North London Science Centre and directed the BBC *Primary Science* and *Bath Science 5 to 16* curriculum projects.

Liz Lander is a senior lecturer in science education at Roehampton Institute London, working on initial teacher education courses as well as INSET. Liz has worked as an advisory teacher for Surrey local education authority providing in-service courses for teachers, and also as a freelance consultant providing in-service packages for schools.

Barbara Parson is a senior lecturer in science education at Roehampton Institute London, working on initial teacher education courses, both PGCE and undergraduate, as well as INSET. She has previously worked as an advisory teacher for ILEA providing in-service courses for teachers.

Virginia Whitby is currently the assistant TSA coordinator for science education at Roehampton Institute London. She is involved in teaching across all programmes including INSET. Her past research interests have included work on teachers' questioning skills, and their attitudes to science. She is currently involved in research to examine the effect that teachers' own scientific knowledge and understanding have on their science teaching.

Maggie Williams is a freelance consultant for primary science. She currently works on initial teacher education courses at Roehampton Institute London. She has previously worked as an advisory teacher and a head teacher within the ILEA. She has wide research interests into children's learning in science.

Introduction

This book has been written to enable students in initial teacher education and newly qualified teachers to make effective and informed links between scientific knowledge and its application for children in the primary school classroom. It provides an introduction to the nature and content of science, how children learn science, and how teachers can support scientific learning. It has been organised for ease of reference to the requirements of the science curricula of the United Kingdom, and will give readers a broad pedagogical base for science.

CHILDREN'S LEARNING

Within this book we have approached science through the constructivist view of learning. Constructivism is a perception of the way learning takes place. Learning is an active process involving the selection and integration of information by the learner. It is very much the opposite of the view that children are empty vessels to be filled with knowledge.

Through recent research projects such as 'SPACE' (Science Process and Concept Exploration Project 1991), and 'CLISP' (Children's Learning in Science Project 1984–91), it has been shown that children hold many ideas about how the world around them works. Children therefore develop their own scientific ideas and formulate their own understanding whether or not they are formally taught science. This can lead to children holding problematic misconceptions which can hinder their scientific learning. It is essential that teachers find out what children know at the outset of learning, to establish their understanding of science concepts. This constructivist view of teaching and learning is promoted by both the SPACE and CLISP projects, and we shall refer to both projects throughout this book.

Both children and adults construct, or build up, their own concepts. A concept is made up of the inter-relation in our minds of facts, ideas and understanding. Only when these three aspects are joined can we construct concepts to explain what we see and experience. Concepts will always be rooted in our own previous experience, which will influence our attempts at understanding. There is a sense in which our concepts can never be 'wrong', as they are a reflection of our own level of understanding at that time. They may not, however, concur with accepted scientific ideas and may therefore be 'misconceptions'. For us to be able to alter our conceptual understanding, new experiences must be encountered which challenge the existing concept. If this challenge involves too big a leap from currently held ideas, the new ideas will be dismissed out of hand, or lip service will be paid to them while the original concept remains intact. If the challenge involves a sufficiently different rethink of the currently held concept, then the concept can be adapted or changed to fit the new experience; learning will therefore take place. A major part of our role as teachers must therefore be to ascertain children's current understanding and then structure new experiences to bring about this learning.

CONCEPTS AND CONTEXTS

Science for both adults and children is about making sense of the world around us and how it works. What many of us appear to have inherited from our secondary school experiences of science is isolated chunks of knowledge or individual facts and definitions which we are unable to link together or apply even if we are able to recognise situations where they might be relevant. What we have is 'knowledge', often without 'understanding'. The aim of science is to help us understand the way things behave and to develop broad conceptual understandings which we can interlink and which help us to find patterns in phenomena and the behaviour of our world.

As teachers we are aware of the need to find effective ways of establishing the ideas that children bring to their learning in science. An idea known as concept mapping was developed many years ago and this has proved to be a very helpful tool for both children and teachers in eliciting children's own understanding.

To introduce the idea of concept maps teachers need to have a discussion with the children to establish their own ideas. As a result of the discussion the teacher can provide an appropriate set of words for children to use. For example, the teacher gives a concept map to the children; they could write the word 'trees' and draw a circle round it and then write the word 'leaves' and do the same. They could then join the two circles with an arrow pointing towards 'leaves'. What then could be written on the arrow to describe a relationship between trees and leaves? It could be 'grows', 'reproduce', 'feed' or, 'drops'. Each of these responses would indicate a slightly different understanding of the relationship between trees and leaves. Each of these words could then be linked to other words concerning trees and leaves and the result would constitute a concept map. Examples of concept maps for materials and electricity can be found in Chapters 3 and 4. We recommend that you try a concept map for yourself before planning a topic, and complete a further one at the end. It may be interesting to note any developments or changes in your own scientific understanding through researching and planning the topic.

The purpose of primary science therefore is to equip children with the skills to tackle new situations with confidence and to develop broad conceptual knowledge and understanding as a solid base on which to develop new concepts.

SCIENCE AND LANGUAGE

The following example of scientific thinking occurred when working with a reception class and introducing them to floating and sinking. The children had a tray of objects and were taking turns to select one object, feel it, and then predict what would happen when it was placed in the water. One girl aged 4 years 10 months chose a block of wood.

Teacher What do you think will happen when you put it into the water?
Child It will stay on the top not moving.
Teacher Will it float?
Child No it will stay still on the top.

The child places the wood into the water and it is floating.

Teacher Is it floating?
Child No it is staying still.

At this point it was obvious the child was not able to accept the term floating in relation to the tank of water and the wooden block. The teacher therefore needed to find out what the term floating meant to her.

Teacher What is floating?

The child looked at the teacher very excitedly and said 'Floating is when you are in the swimming pool and you have a float and you are moving around'.

The child had a clear understanding of the term floating that she was confident with. For her it involved movement, but she was not yet ready to take on floating in relation to the block of wood and the water tank.

This example makes it clear that language plays an important role in developing children's scientific thinking. Language is the vehicle through which we communicate. The importance of establishing children's own ideas and understanding is essential so that we can then provide them with appropriate science. It also allows children to develop their language skills in a variety of contexts. You will find further evidence of the role of language in science throughout each chapter of the book, where we encourage children to talk to their teachers and, more importantly, to each other about their work.

MATHEMATICS AND SCIENCE

There are many natural links between science and mathematics. Much of science involves aspects of measurement, sorting and comparing, and classifying. Both maths and science provide opportunities for predicting skills to be developed. Both subjects also require the learner to make appropriate choices about the line of action to take, and encourage the learner to use alternative strategies when their first choices prove unworkable.

SCIENCE IN THE NATIONAL CURRICULUM

Since the introduction of the National Curriculum in 1989, science has been a core subject. However, science has been a cause of concern for many schools, as before 1989 it did not have the same status in schools as the other core subjects of English and mathematics.

The National Curriculum for science is composed of four areas, all of which are addressed in this book. The first area is 'Experimental and investigative science', which is the process of how children learn to behave scientifically, in other words to develop the ability, for example, to observe, predict, obtain evidence and design fair tests. Whatever the content area of science children are engaged in they will be using and hopefully developing their scientific skills. 'Life processes and living things' is the biological area of the National Curriculum, 'Materials and their properties' covers the chemistry and 'Physical processes' the physics.

The demands of the National Curriculum for science at Key Stages 1 and 2 require a substantial understanding of scientific concepts, to be built on and developed throughout Key Stages 3 and 4. Since its inception, many primary school teachers have raised concerns about their own level of scientific knowledge and understanding, as many of those teaching prior to the National Curriculum received no science teaching during their own initial teacher education courses.

However, science is now an important aspect of all initial teacher education courses alongside English and mathematics and the foundation subjects. From September 1998 the Teacher Training Agency will set out criteria for student teachers to ensure that their scientific knowledge is of an acceptable standard. OFSTED (1996) state that there is a strong belief that knowledge and understanding of a subject facilitates more effective teaching and learning. There is currently debate about the most effective way to approach the science curriculum. Some schools favour a subject-specialist approach particularly in Years 5 and 6 as this ensures that the content is covered effectively. Other schools find the traditional topic approach more comfortable as learning is not so compartmentalised. Science does pose particular problems, but it highlights a broader debate about the role of subject knowledge in teaching.

The first four chapters of the book address the demands of the National Curriculum for primary schools and desirable learning outcomes for early years units. Each chapter includes science content, children's ideas, teaching advice and case studies. Set out below is a brief synopsis of each chapter.

- Chapter 1 explores the nature of experimental and investigative science; that is the process skills of science. It shows how children and scientists use these skills, and their implications for classroom practice. This chapter discusses the nature and place of practical work, the importance of context and the progressive development of these process skills.
- Chapter 2 explores life processes and living things, thus focusing on the biological aspects of science. It considers the scope and variety of life, and presents ways in which children can develop their understanding of these and gain respect for the natural environment.
- Chapter 3 shows how development in the understanding of the behaviour of materials can help to support scientific investigation in all other areas. Exploring materials provides stimulating first-hand experience for children across Key Stages 1 and 2, enabling them to make sense of their environment. The areas included in the chapter are: solids, liquids and gases; heating and cooling;

mixing and separating; chemical changes; mechanical properties; obtaining, making and using materials.
- Chapter 4 considers the physical processes of science. This is an area which is more abstract and difficult for both children and adults to understand. The concepts of electricity, energy, force and motion, light and sound, and the Earth in space are all covered in an accessible way, and methods of developing children's understanding are detailed.
- Chapter 5 considers the planning and the development of a scheme of work. In addition issues concerned with classroom organisation, appropriate resource provision and resource management will be discussed. Methods of recording and assessing science in the primary school are also discussed in this chapter.

We hope that as readers you will find this book provides you with the opportunity to develop your own scientific knowledge and understanding. In addition we hope that the practical advice and case studies will enable you to find ways of implementing the National Curriculum for science in a lively and exciting way for the children who will be the scientists of the future.

CHAPTER 1

Experimental and Investigative Science

Science is often thought of as certain kinds of information, for example 'How dinosaurs lived, and why they became extinct'. It can equally well be characterised as a process of enquiry – how we find out about dinosaurs rather than what we know about them. In this chapter we will explore this process, usually called 'investigating', as it is practised by children and by scientists, and how it is to be implemented in schools to meet the requirements of the curriculum. Firstly, let us analyse an enquiry by a child (Figure 1.1).

Figure 1.1 The process of science investigation

THE PROCESS SKILLS OF SCIENCE

John asked the question: 'Why is Lisa's dress blue?' This is prompted by an observation that it *is* blue, and more importantly, that it is a matter of interest. What kind of answer is expected to this question? 'Because Lisa is wearing the school uniform' will not result in any further enquiry, so we need to elicit from John why he finds this question interesting. It may be that he has done some work on mixing paints to make different colours and wonders if the same thing happens with cloth. He may have noticed that clothing seems to change colour under disco lighting. After exploratory activity and discussion, John decided to focus on the question of coloured light falling on coloured cloth. He was able to formulate a prediction: 'I think that when I shine green light on the blue cloth it will look greeny-blue'. This is related to a more general hypothesis: 'Coloured lights have the same effect as coloured paints'.

This can now be tested, but he needs to plan how the test can be made fair. Does he really need Lisa, or only her dress, or would any sample of blue material be suitable? How should he arrange to shine the green light? Does all other light need to be excluded? How bright is the light to be, and what shade of green? All these factors could affect the result. They are variables to be investigated or controlled in the experiment.

The observations or measurements he makes provide the evidence; it is important to record this carefully so that it can be correctly interpreted. For example, it will probably make a difference to his results if the colour of the light is different shades of green and the colour of the fabric is different shades of blue. Unless this is systematically recorded then the wrong interpretation could be made. The final step is to compare the results with the prediction and draw a conclusion. In this case, if the light was pure green and the fabric was pure blue then the fabric's appearance would be black. This would not confirm the hypothesis. If however, as is more likely, neither colour is pure then John would get the result he expected and his hypothesis would be confirmed.

This seems disturbing – how can we accept the possibility of conflicting conclusions from the same investigation? This is the reality of the situation, however, and it points to three conclusions about the nature of the scientific process. The first is that repetition is important. Anyone could confirm John's results with the same lamp and the same fabric. The second is that evaluation and communication must be an essential part. The contributions of several investigators can then be combined to perhaps discover that it is the variations in the shade of the colour that give these apparently conflicting results. Finally, the conclusions of the enquiries should always be considered as *provisional* knowledge. In this case John's ideas could be replaced by a better hypothesis or theory of colour mixing, as described in Chapter 4 of this book.

Whatever facts or theories come out of scientific enquiry, the processes are always similar. These are the 'doing words' given in **bold**. They can be grouped into three clusters:

- **planning** including formulating questions from significant observations, making predictions and hypotheses which can be tested, planning fair tests in which variables are controlled;
- **doing** including selecting and using equipment, observing and measuring to collect evidence, recording data;
- **interpreting** including organising and presenting results, concluding with respect to predictions or hypotheses, evaluating experimental work and findings, communicating to others.

These could be seen as the before, during and after stages of the experiment, but often there are trial and exploratory stages, so the whole enquiry has more of a cyclic nature (Figure 1.2).

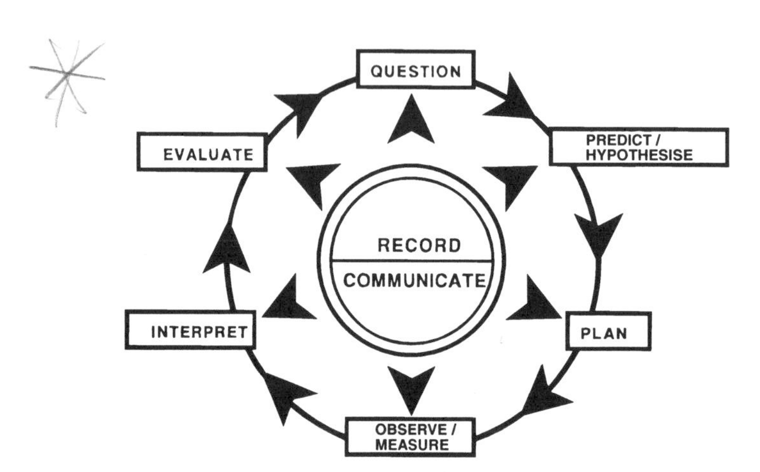

Figure 1.2 The skills of science investigation

SEX, DRUGS, DISASTERS AND THE EXTINCTION OF DINOSAURS

Here is a story from science in which the same processes are operating (Gould 1984).

Dinosaurs, their lifestyles, death, and the possibility of their reincarnation in our own times has fascinated people long before the *Jurassic Park* movie. Evidence of how they lived comes from fossils, as does the evidence of their mass extinction after 100 million years of global domination. Quite simply, their fossil record read from rock dating stops 65 million years ago. The science begins with this **observation**, which may also be expressed as a **question**: What happened 65 million years ago? This might be thought rather hard to answer, but, perhaps surprisingly, several explanations have been offered by scientists in recent decades.

These are the hypotheses:

- **sex**: there was a global warming at that time; dinosaurs being large were unable to cool their bodies sufficiently, the males' testes got overheated and they became infertile;
- **drugs**: flowering plants evolved at that time; many of these contain toxic

substances to protect themselves from being eaten, dinosaurs didn't recognise this and died of poisoning;
- **disasters**: there was an impact of a large asteroid at that time; the debris it created blocked the sunlight and cooled the planet so dramatically that many species died of freezing or starvation.

These hypotheses need to be tested; investigations require planning so that evidence can be collected. How is this done? The only source of evidence for events so long ago is found in the rocks and other deposits that were formed at the time. Testes do not fossilise so it is impossible to know whether the sperm were infertile, so the sex hypothesis cannot be tested. For the drugs hypothesis, there is fossil evidence that many dinosaur bodies were distorted in death, which could suggest that they died in agony from poisoning. However, there is a simpler interpretation – the distortion was caused by muscle contraction after death, and by movement of the rocks later. Nor can the drugs idea explain why many other species became extinct, or why dinosaur extinction happened tens of millions of years later than the evolution of flowering plants. Evidence for the disaster hypothesis is from deposits of unusual materials at that time, as would be expected from the collision of an asteroid. Other scientists argue that these materials could have come from within the Earth as a result of volcanic activity. Their evaluation of the evidence does not destroy the idea completely; there would still be a dust cloud blocking the Sun, but its source has been modified.

This 'atmospheric disaster' hypothesis is currently the best theory of how the dinosaurs, and many other species, became extinct 65 million years ago, even though there is no general agreement on all the details. The theory evolved by scientists communicating; more significantly, they discovered something entirely new as a result. This is even more important to humans than the extinction of the dinosaurs – the possible extinction of the human species! It was previously thought (by those who can contemplate 'megadeaths') that a nuclear war was winnable. It would take a while for radiation to fall in bombed areas, but then they could be invaded. When this 'atmospheric disaster theory' was applied to nuclear war, it showed that the whole Earth could become uninhabitable because of the debris and smoke from fires blocking the Sun for years. The influence of this theory is believed to have helped end the nuclear arms race between the USA and USSR.

This story makes the case that the process of investigation is more important than the facts it establishes – we still don't know for sure what killed off the dinosaurs, but we may have avoided killing ourselves off.

KNOWLEDGE, UNDERSTANDING AND SKILLS IN SCHOOL SCIENCE

For school science there has been considerable debate over many years about the relative importance of the content of science – facts and theories to be known and understood; and the process of science – skills and procedures to be practised and used. Defining the content can be important because:

- it is impossible to cope with all possible enquiries that children may have, so there needs to be some selection;
- certain knowledge and understanding is particularly fundamental in order to understand the world;
- certain knowledge and understanding is especially useful, and is motivating for children;
- children don't know which knowledge is the most important;
- teachers need help in providing a balanced content;
- schools need help to cover suitable content at the best time and to prevent omissions and duplications;
- an agreed content results in a more valid assessment of pupils' learning.

The arguments for emphasising processes include:

- what is considered the most important knowledge and understanding will change with time, because that is the nature of science;
- what is useful and motivating to children will depend on the children concerned and will vary;
- specifying the content inhibits the way in which the teaching is organised, e.g. making topic approaches more difficult;
- these skills help children to behave in a scientific way;
- the skills and procedures of investigating can be applied to new content and make it easier to understand;
- a process approach emphasises the need for active engagement by the learner, and this is a more effective way of developing understanding.

The requirements of the curriculum for science in primary schools reflect this debate. Content and process are both specified, and are supposed to be given equal weighting in the planning of activities – and in the assessment of pupils. However, there is also an emphasis on the relationship between content and process: the skills have to be practised and used on some scientific content, and the knowledge and understanding will be gained through the use of a skill. For example, CD–ROMs can be useful stores of scientific knowledge and understanding, but one needs to learn information technology access skills to make use of them. This book represents a body of science content grouped around a major theme in each chapter. The writers assume that the readers will have the study skills to learn some science from this content, but also hope that such learning will be supplemented with activities which develop the readers' scientific skills as well!

HOW CAN CHILDREN DEVELOP THE SKILLS OF SCIENCE?

It is appropriate to consider children *developing* these skills rather than learning them, as it can be argued that children have been using them to make sense of their world from a very early age. Research in recent years has produced some remarkable findings, through ingeniously devised experiments, such as mobiles which can be controlled by babies' hands, or lights which respond to their head turning. Babies of only a few months old have been found to explore a world that they conceive to be 'out there' and to be able to affect this world in ways that they choose. Researchers think that this process can have important consequences for their later development (Donaldson 1992).

By the time that children start nursery school they will have had considerable experience of interacting with the world in an investigative way. The role of the teacher, therefore, is that of developing these skills by selecting scientific contexts which draw on children's personal experiences and challenge them to progress further. The statements of curricular requirements chart the directions of that progress. For example, in the National Curriculum for science in England and Wales, five levels of attainment are identified, each representing, in broad terms, a year or two of a child's school career. This aspect of science is called **Experimental and Investigative Science**, and is divided into strands called 'planning experimental work', 'obtaining evidence' and 'considering evidence'. This is similar to the before, during and after sequence described in the section on 'The process skills of science' above. The way these skills are seen as developing is shown in Figure 1.3. The processes of science are given slightly different labels in the curricula for Scotland and Northern Ireland, but they each describe similar progressions.

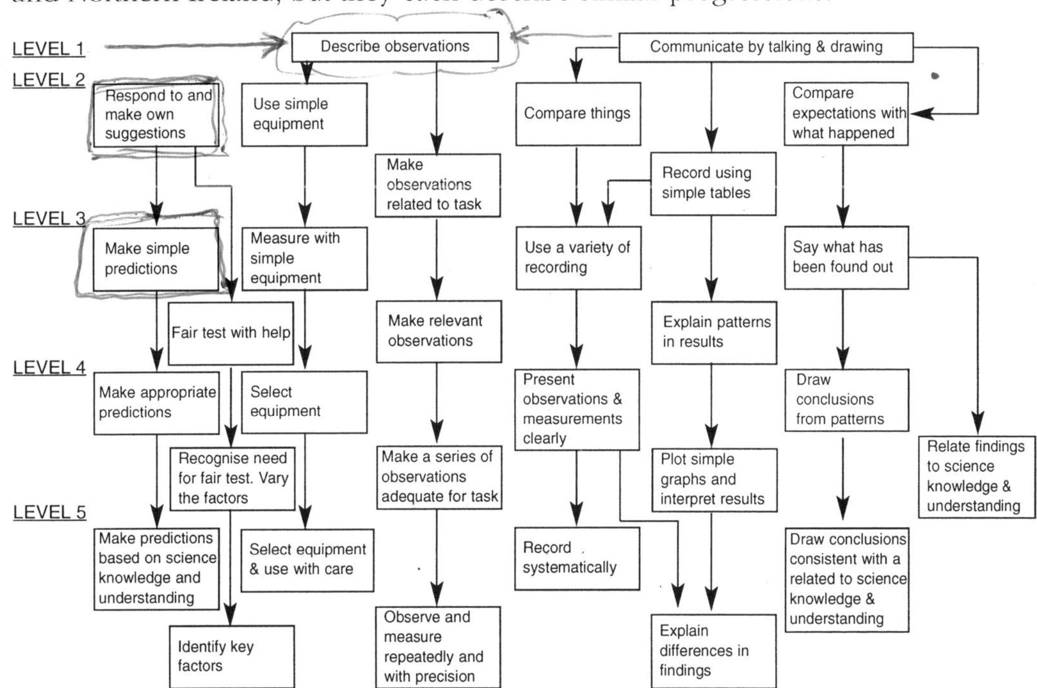

Figure 1.3 Progression in experimental and investigative science (Science 1)

Practical, hands-on experience is clearly essential in developing these skills. Indeed 'good' science teaching is often seen as synonymous with practical activity. In reality things are a little more complex. Practical work is not necessarily investigative, and investigating is not always 'hands-on'. A useful analysis of types of practical work was published by the (now defunct) National Curriculum Council, to help teachers interpret this aspect of the recently introduced National Curriculum. Four distinct purposes were identified:

1. development of basic skills – e.g. practice in using a thermometer or microscope;
2. illustration of a principle, idea or concept – e.g. that water boils at 100°C, that day and night are caused by the rotation of the Earth;
3. observation – the differences between two flowers or two local habitats, the speed of a toy on a ramp;
4. investigation – a full enquiry in which the practical work is preceded by questioning, predicting and planning and is followed by interpreting, evaluating and communicating.

Each of these has a valid place in the teaching of science, but only the last one meets all the requirements of 'Experimental and Investigative Science'.

It may be noted at this point that there is a common misconception about the word 'experiment'. It is often used in the form 'an experiment to prove that pure water boils at 100°C'. This is an example of point 2; not an experiment but an illustration of a principle – science has *defined* that pure water boils at this temperature. As part of an investigation into water supplies one might predict that a given sample of water is pure and carry out an experiment to test the prediction, i.e. if it does boil at 100°C then it is proven to be pure. An experiment is a test of a prediction which will usually be part of an investigation.

The other side of investigations is that there are processes which are 'minds-on' rather than 'hands-on', as noted in point 4. These are often the parts that children find more difficult. They are keen to get into the 'experiments' before they have really thought about why they are carrying them out, and whether their findings will be useful. It is often helpful for the teacher to suggest a series of explorations or 'trial-runs', which are then thought about, as an aid to planning the full test or experiment.

INVESTIGATIONS IN THE CLASSROOM

It might be thought that the way for a teacher to get good investigations in their classes is to let the children get on with it alone. However, the progression identified in the curriculum requirements (Figure 1.3) emphasises that pupils need to develop these skills. The teacher's role, therefore, is to intervene appropriately to facilitate this development. This will usually mean more intervention with younger children, or in less familiar contexts.

Providing a suitable context is the first important role for the teacher. This will arise naturally out of the study of a scientific or related topic, e.g. the behaviour of paper in wrapping a parcel. The teacher then needs to decide what the scope of the enquiry will be, and the particular skills that will be developed by it. With younger children the focus might be the identification of paper as a material. The activity could be sorting into 'paper' and 'not paper', so that the skill is observation and in particular comparison and identification of difference. This aspect could be extended for older children with the provision of magnifiers and microscopes for more detailed observations which can be linked with ideas on the origins and properties of the different types of paper. This could then develop into a whole investigation, with the pupils being encouraged to formulate questions and predictions based on these observations and their prior experience, e.g. they may choose to investigate strength or 'waterproofness'.

Pupils often need most help in the planning stage of an investigation. First of all they need a prediction to test, and the teacher can help refine this with a range of open-ended questions such as 'What do you think will happen if . . . ?, Would it make any difference if . . . ?, How can you be sure that . . . ?' and perhaps most the important 'Why do you think that . . . ?' which focuses on the child's own ideas and tries to avoid the response of giving the teacher what the child thinks she wants to hear. It confirms the approach that it is the child's ideas that are important at this stage, not some expected right answer. The aim is to develop children's abilities to ask the right kind of questions themselves. It can be argued that this is the most important step in the whole process. It is the kind of question that is asked which defines the enquiry as scientific. It should direct the enquirer into thinking about how to plan a way to find an answer, and also about what it is they already know, which makes this a question *worth* investigating.

There has been considerable research on the best way to do this, e.g. see Watts *et al.* (1997) and Gilbert and Qualter (1996). This exploratory planning part of the process can be carried out with a whole class and the ideas considered by all and refined into an experiment, or a series of investigations, each carried out by small groups. The idea of a fair test is crucial in science, and can be introduced with reference to games that children play – they usually have a very well-developed sense of what is fair in this context! The identification and control of variables, which will follow planning for older pupils, can be supported by listing them on a sheet of paper and perhaps cutting the list into pieces for each group to control one and measure another.

During the practical work pupils will need help in a range of different ways. Probably the most important factor in determining how to organise the class for investigations is the number of children the teacher can help at the same time. Failure to anticipate their demands could lead to the following scenario involving a teacher inexperienced in science:

I went into a Year 2 class to help the teacher and was faced with 12 unrelated activities set out around the classroom. This was also an unfamiliar situation for the children, who as a result were 'high'. The class teacher and I spent the entire afternoon racing around like headless chickens trying to help the children as they demanded it. At the end of the day as we sank down exhausted she said: 'See I can do it, I can teach primary science!' I was left wondering what we had actually taught the children and why for this teacher 'primary science' seemed to imply juggling an impossibly large number of activities.

Even with a narrow focus to the activity there are many demands. Younger pupils may need help with manipulative skills, e.g. perhaps with paper that needs cutting. They also need to be kept on task by prompting with questions and reminders of the purpose of the activity. Older pupils will also need prompting to check that they have considered all the relevant factors and that they are controlling variables to ensure a fair test. For example, is the way that they test the strength of paper, say by tearing, both valid and reliable?

Pupils will always need teacher intervention to help them record and interpret their results – it's much more fun to go on investigating! For young children, recording using the actual materials, e.g. sticking the group of papers up on a poster, has the advantage of also providing a means of communication which others can evaluate, i.e. do they agree with this grouping? Older pupils need help in sorting out their conclusions from factors they didn't cover, e.g. 'This paper is the strongest covering for parcels, so long as it keeps dry'. The limitations of any one conclusion can be the spur for further work, either by the group concerned or by a different group who can check and extend the enquiry.

Any practical activity should end with all the participants feeling that they have learnt something. In the bustle of activity, children do not always think about this. In the rush to clear up at the end of a session, teachers do not always find time to remind children about the purpose of the activity. It is important, therefore, that at the next opportunity teachers help pupils to reflect on what they have learnt – whether it was a skill such as using a magnifier, the ability to pose an interesting question, or the discovery of an answer through using some of their prior understanding of scientific principles.

CHAPTER 2

Life Processes and Living Things

INTRODUCTION

Life processes and living things is an area which provides a constant source of fascination for children as they explore the natural world around them. Investigating the living world provides an enormous range of opportunities for developing investigative skills in science, fuelled by children's natural curiosity and their need to make sense of the world around them.

This chapter looks at this wealth of material from two approaches:

1. characteristics of living things;
2. environment.

Although we are presenting the chapter as two main areas of exploration, this separation risks loss of meaning. To maintain a coherent whole within this very wide-ranging area, we can identify several main strands which represent the major concepts encountered during any investigation of living things. These concepts will provide useful signposts to refer to throughout the chapter. Misconceptions can easily arise when living things are studied in isolation without taking these strands into consideration. For example, a study of seed growth may involve opportunities to observe and record aspects of the life cycle and patterns of growth. Using the seven strands in Figure 2.1, a fuller investigation could be achieved, as shown in Table 2.1.

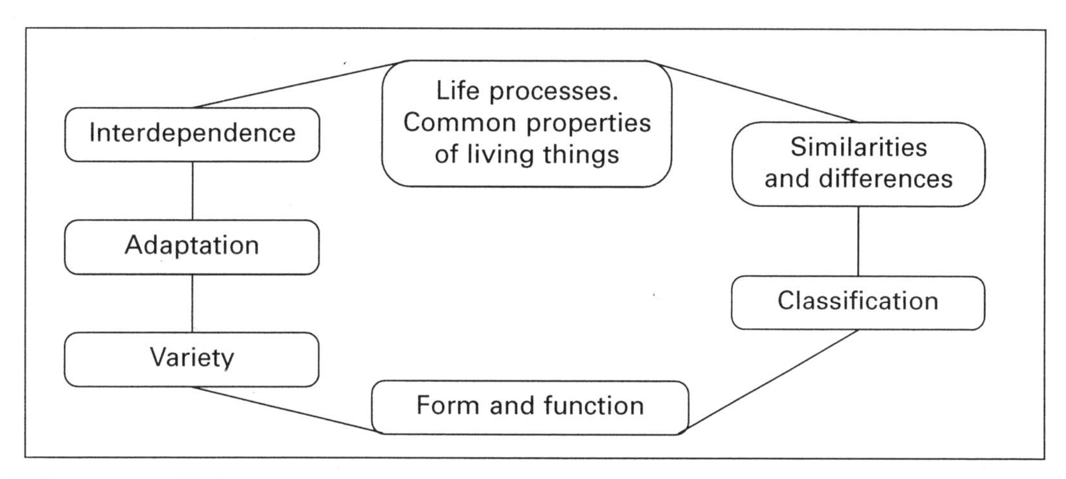

Figure 2.1 Major concepts in living things

Table 2.1 Investigating broad bean seeds

Similarities and differences	Compare and contrast with other types of seeds: colour, shape, size, weight, etc.
Variety	Look at various types of seeds to note variety of form.
Form and function	Observe and identify parts of a seed. What role do these parts play? Observe form and function of parts of mature plant.
Life processes	E.g. growth, life cycle. Where did seed come from? Sensitivity of plant to light. Needs of seed and plant for growth and development.
Adaptation	How is seed dispersed? What mechanism is used? In a mature plant, how is pollination achieved? How is flower adapted to attract bees? How is the plant suited to its environment?
Interdependence	Need for insects for pollination. Broad bean as a source of food. Role in food chain.
Classification	Identifying details to decide on grouping of plant. Can you find other plants with similar flower structure?

THE BASIC UNIT OF LIFE – THE LIVING CELL

To have an understanding of life processes, it is useful to know something about the units which make up living material and the way in which these units combine to form larger structures responsible for carrying out specific functions. The basic unit is the cell. There is no such thing as a typical plant or animal cell, because they can vary in shape or size depending on their function, although all cells share certain characteristics. Figure 2.2 highlights the similarities and differences between plant and animal cells.

Humans are made of millions of cells, each cell carrying out essential life processes. As animals increase in complexity from the single-celled *Amoeba*, cells capable of a specific function become grouped together to form organs. The nervous system, heart, liver, kidneys and bones are all organs made up of the individual units of life. The diversity of cell type found within the plant and animal world is enormous (see Figure 2.3) and makes us very aware of the wide variety of form and function.

All cells need similar chemical compounds to maintain essential life functions although the amounts needed may differ. Table 2.2 lists these essential substances and how they are used in the living cell.

LIFE PROCESSES

Understanding life processes

Life processes are a group of functions common to all living things. The way in which they appear to be carried out may differ, but these functions are essential for maintaining life.

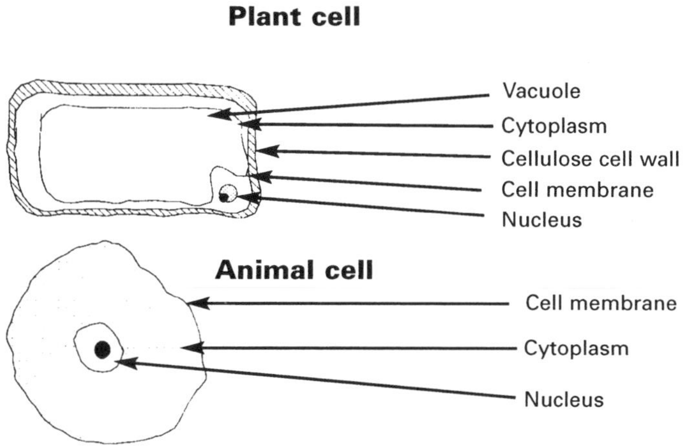

Plant cell

Vacuole
Cytoplasm
Cellulose cell wall
Cell membrane
Nucleus

Animal cell

Cell membrane
Cytoplasm
Nucleus

Similarities

(a) All cells have a cell membrane surrounding them which separates the inside from the outside of the cell.

(b) All cells have a nucleus or nuclear material where the genetic material is stored in the form of deoxyribose nucleic acid (D.N.A.). The DNA controls chemical activity in the cell and is responsible for hereditary information.

(c) All cells contain an inner liquid-like medium called cytoplasm, where all chemical reactions take place e.g. respiration.

Differences

(a) Plants have a non-living cellulose cell wall outside the cell membrane.

(b) Most mature plant cells have a central fluid-filled space called a vacuole. The vacuole contains a mixture of water, salts and sugar and plays a central part in maintaining plant shape and structure. Water loss results in wilting.

(c) The cells of green plants contain the green pigment chlorophyll which is responsible for absorbing the energy from sunlight needed for photosynthesis. Chlorophyll is found within particular structures called chloroplasts.

Figure 2.2 Typical plant and animal cell showing general features and similarities and differences

Basic life processes comprise:

- Nutrition
- Respiration
- Movement
- Excretion
- Reproduction
- Growth
- Sensitivity

Nutrition

Looking at the list of materials needed by all living cells gives us some idea of the importance of nutrition (Table 2.2). All living organisms need food as a source of raw materials

White blood cell (neutrophil)

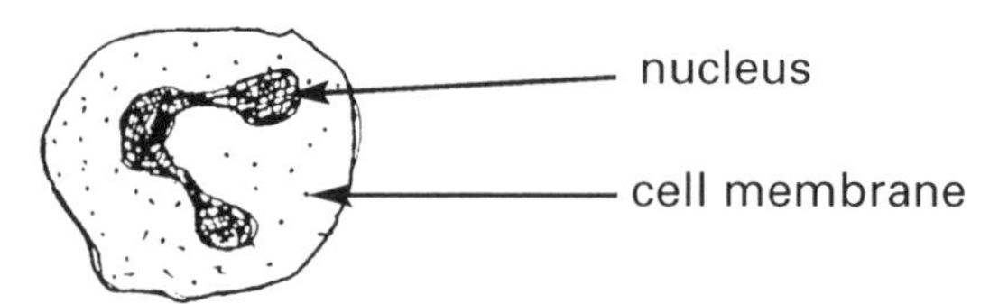

These cells play an important part in protecting the body against infection

Motor nerve cell

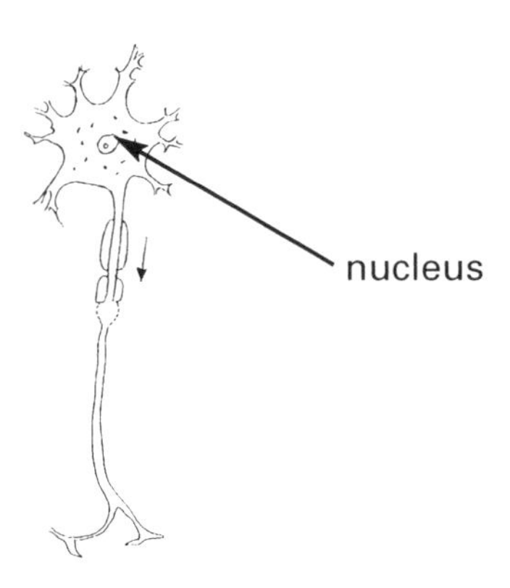

Palisade cell found within leaf of green plant

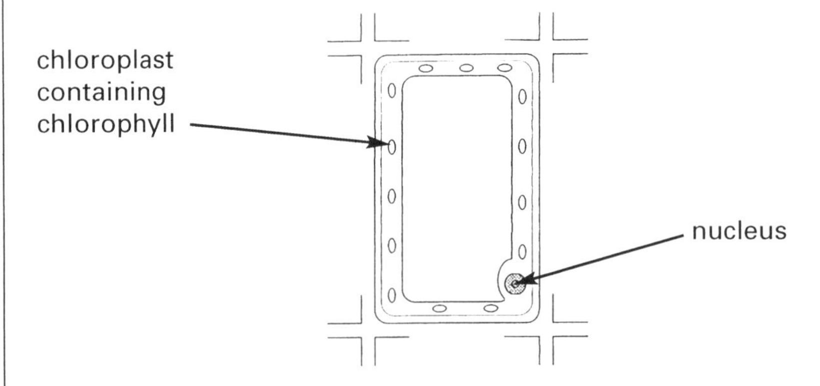

The palisade cells form a layer in the middle of the leaf. The pigment chlorophyll absorbs energy from sunlight which enables photosynthesis to take place. The elongated sides of these cells ensure that the chloroplasts get maximum exposure to sunlight.

Figure 2.3 A range of specialised cells in plants and animals: white blood cell (neutrophil), motor nerve cell and leaf cell (palisade cell)

Table 2.2 What cells are made of

Substance	*Function*
Water (75%)	Allows things to dissolve, move about and keeps cells from collapsing.
Proteins	Needed for cell structure, used in chemical reactions. Made up of carbon, hydrogen, oxygen and sulphur.
Lipids (fats and oils)	Needed for cell structure and can be a source of stored food for energy release, e.g. fats in seeds. Made up of carbon, hydrogen and oxygen.
Carbohydrates	Needed for structure, e.g. cellulose. Commonest form is glucose. Starch is a storage form. Made up of carbon, hydrogen and oxygen.
Salts	Needed as part of chemical reactions. Also used in the transmission of electrical impulses in nerve cells.
Vitamins	Needed in chemical reactions. Plants can make their own vitamins.

for building and maintaining cell structure, for growth, and essentially for providing energy to carry out particular functions.

The primary source of all food is the green plant. Green plants are able to produce carbohydrates using carbon dioxide from the air, together with water absorbed from the soil by the root. This process, known as photosynthesis, uses the energy from sunlight to build up sugars from simple substances (Figure 2.4). These sugars are the building blocks for other carbohydrates, fats and proteins.

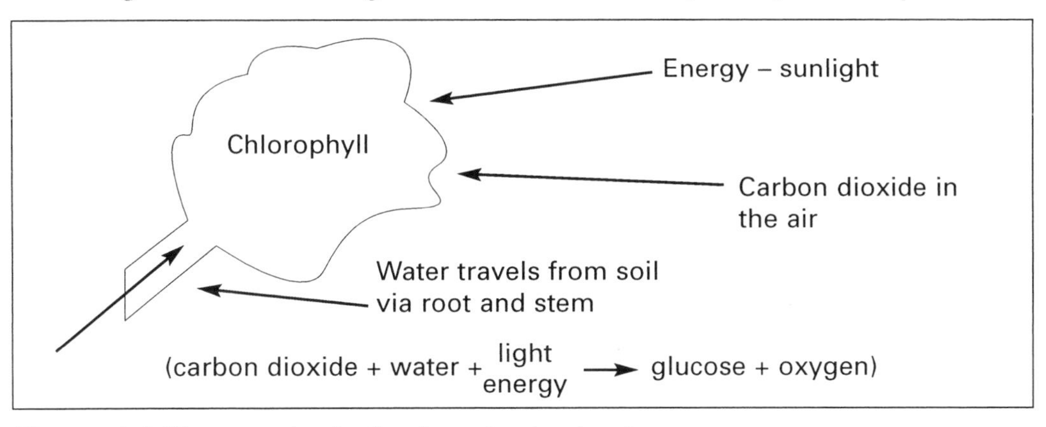

Figure 2.4 Photosynthesis: food production in plants

These substances then become the food of animals. For example, humans eat various parts of the plants:

- fruits and seeds: apple, plum, wheat, corn, peas, beans
- leaves: lettuce, cabbage
- stems: celery, rhubarb
- roots: carrots, turnip, swede

Animals that do not eat plants get their food by eating other animals that are plant eaters, so all animal life is either directly or indirectly dependent upon green plants for food. Green plants are called the primary producers. In addition to providing fats, proteins and carbohydrates, plants are also a source of vitamins and minerals.

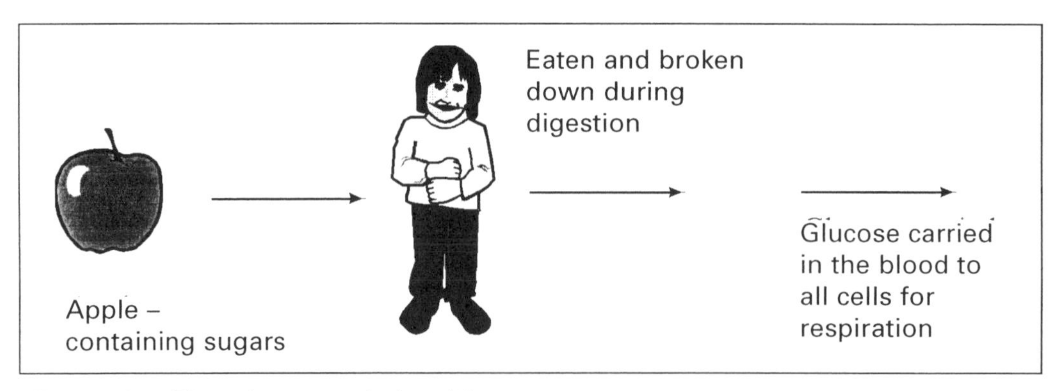

Figure 2.5 Plant–human relationship

Living things need food because the cells of all living things require energy. The energy from the sun is stored chemically in carbohydrates. This energy is released and powers many important functions when the carbohydrates are broken down again during respiration. To illustrate this process, we can take ourselves as an example (see Figure 2.5). Within our own body. we need energy to maintain form and function. When we look at particular cell types, it is probably easier to see where energy is needed to carry out particular work (Table 2.3).

Table 2.3 Work done by types of cell

Cell type	*Function*	*Work done*
Muscle cell	Movement	Contracts and expands
Nerve cell	Sensitivity	Transmits electrical impulses
Red blood cell	Respiration	Absorbs, transports and releases oxygen

Respiration

Respiration is carried out by all living things and is the process whereby the chemical energy of food is released for use in numerous cell functions. Sugars are broken down, usually using oxygen, to release chemical energy, which can be measured in kilojoules(kJ).

glucose + oxygen → carbon dioxide + water + energy

If we compare the process of photosynthesis with respiration, we can see that they are the opposite of each other (Fig 2.6). The green plant is like a factory producing sugars incorporating energy from sunlight, whereas in respiration, the cell breaks down the sugar to release energy and produce carbon dioxide and water as waste products.

If you look at any tins or packets of food, you will see the number of kilojoules that the ingredients will release in respiration. We all have particular energy requirements suited to our needs. Part of maintaining a balanced diet involves taking in food which provides the required amount of energy.

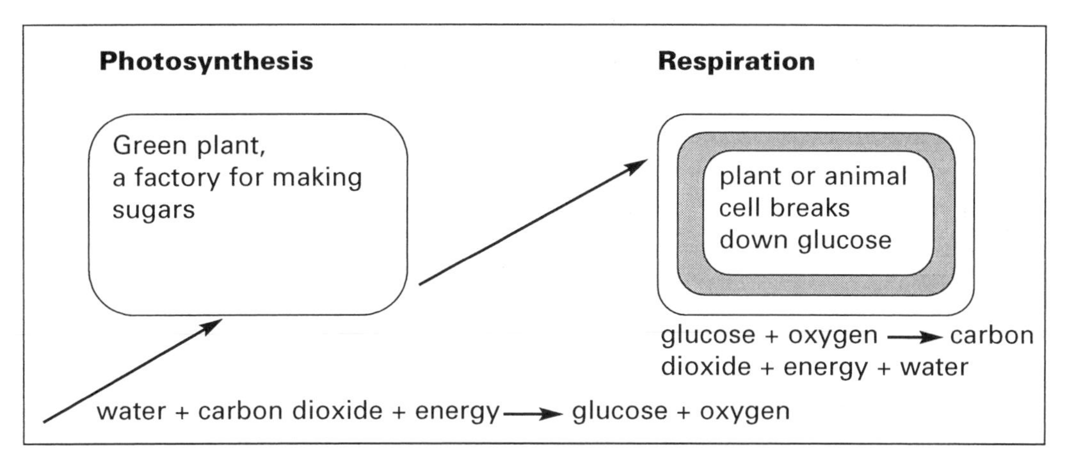

Figure 2.6 Photosynthesis and respiration compared

Breathing or external respiration

We often hear the word respiration used for the act of breathing or *external* respiration, which can cause confusion. Breathing is the method by which oxygen is taken into the body of a plant or animal and then transported to cells to be used in the breakdown of sugars. Breathing also has to remove the waste products of water and carbon dioxide from the cells and then return them either to the air or water surrounding the plant or animal.

Transport systems

To provide the cells of living organisms with the materials essential for life, a system is needed to transport these materials to various sites within the organism. In many groups of animals, blood is the fluid used to transport carbon dioxide, oxygen, and the breakdown products of digestion and absorption, as well as many other substances. In mammals, blood is made up of a liquid called plasma which contains red and white blood cells. Red blood cells contain a substance called haemoglobin, which combines with oxygen in areas where oxygen is plentiful, (i.e. the lungs) and releases oxygen in places where there is a low oxygen concentration (i.e. the body tissues).

The heart acts as a pump to drive the blood around the body, with arteries carrying blood away from the heart and towards the main organs and the veins bringing the blood back to the pumping station. The veins have valves to prevent backflow, and the action of muscles surrounding the arteries and veins helps to maintain blood flow (Figure 2.7).

In plants, transport systems transfer the products of photosynthesis to other sites within the plant where they are needed for respiration, growth or storage. The stem of the plant has specialised cells called phloem which bring about this transport. Special cells called xylem vessels transport water from the roots to the leaves where it is used in photosynthesis.

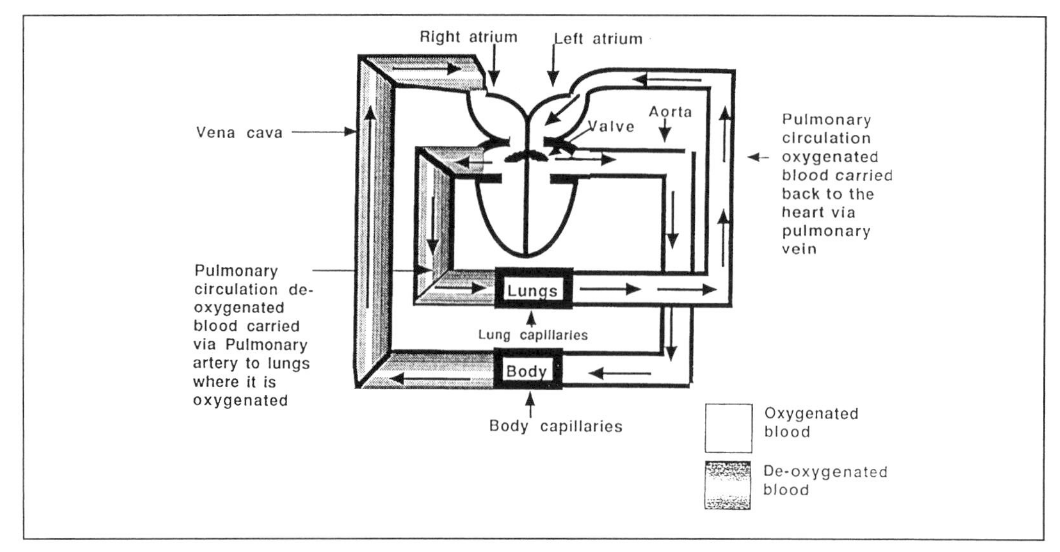

Figure 2.7 Heart, arteries and veins as a functional transport system

Movement (locomotion)

Most animals move about, or at least move parts of their bodies; this movement is often associated with feeding. Not all plants show movement, and when they do it may not be very obvious. Investigations of plants can reveal various plant responses to outside stimuli, i.e. sensitivity. The most important of these outside stimuli are gravity, light, day length, temperature and touch, e.g. the stem and leaves of a plant bending towards the light, the root growing down into the ground in response to gravity, petals closing at night caused by a reduction in light intensity.

Animal support systems and locomotion

Locomotion or movement requires a support system to prevent the organism from collapsing. The skeleton and muscles provide this support in vertebrates (animals with backbones). The skeleton, which is made of bone or a bone-like substance, has special areas for muscle attachment. Other animals, such as insects, have an outer skeleton (exoskeleton) with the muscles attached at particular points to the inside. Both types of skeleton are made of movable parts with flexible areas or joints allowing for movement and locomotion (Figures 2.8 and 2.9).

Many invertebrates (animals without backbones) are able to maintain shape, form and bring about movement through the action of muscles working against an inner fluid. This hydrostatic skeleton is based on the principle that water is non-compressible (Figures 2.10 and 2.11).

The range of support systems within the animal kingdom is very varied (see Table 2.4), but essentially all systems consist of muscles operating in relation to some form of non-compressible structure. We can look at some of these mechanisms within a small range of commonly occurring animals. Skeletal structures show a range of adaptations appropriate for a particular way of life.

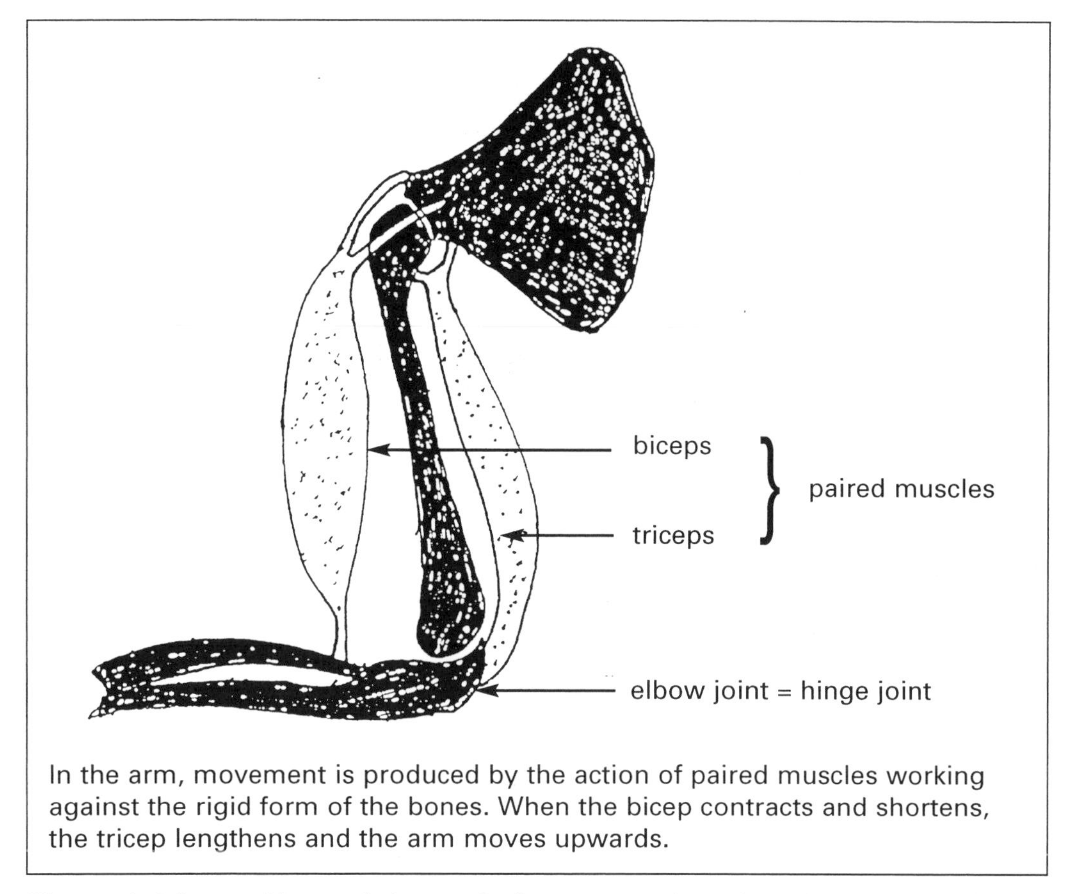

Figure 2.8 Internal bony skeleton of a human arm (vertebrate)

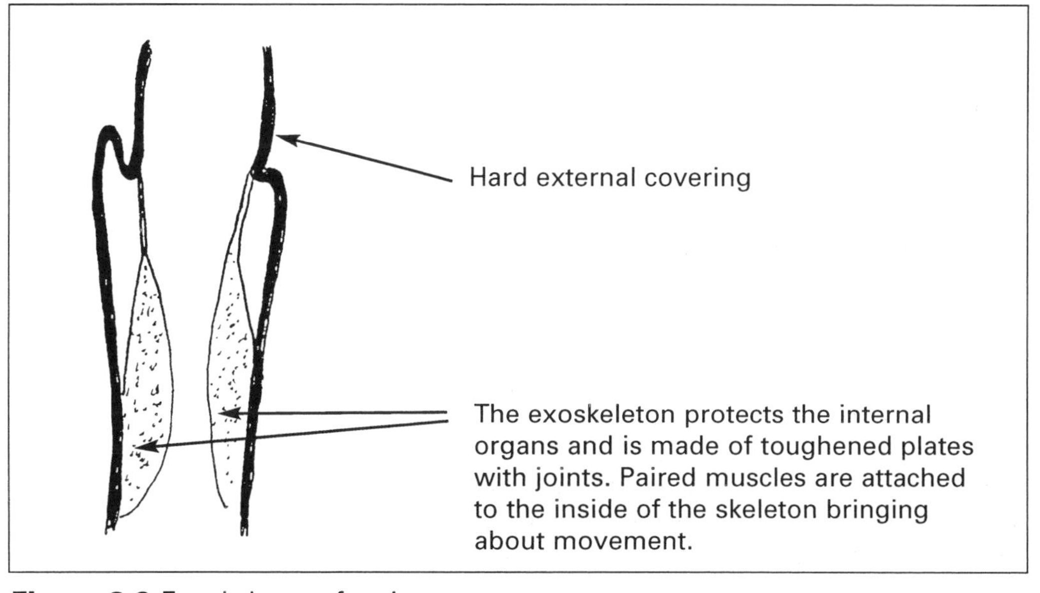

Figure 2.9 Exoskeleton of an insect

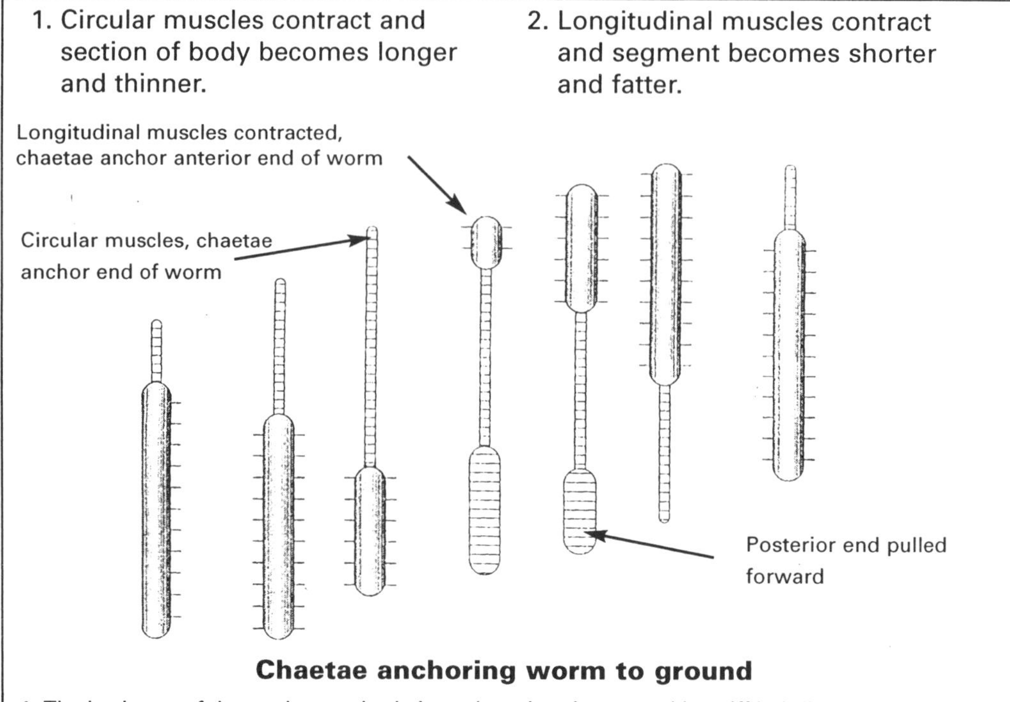

1. The back part of the earthworm body is anchored to the ground by stiff hair-like chaetae while the front section elongates.
2. The front part of the body becomes shorter and fatter by the action of circular muscles, anchors itself to the ground pulling the rest of the body forward. This continues as a series of waves which causes the worm to move forward.

Figure 2.10 Earthworm locomotion

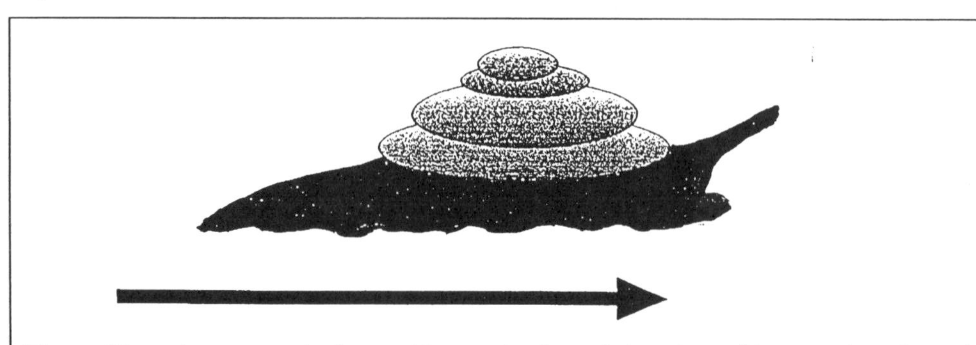

The snail has a large muscular foot and locomotion is carried out by rapid contraction of muscles within this.

If a snail is observed moving on clear plastic, a series of dark and light ripples can be seen moving along the underside of the foot.

The muscles act against the blood filled area within the snail's body which acts as an hydrostatic skeleton. Near the mouth, slime is produced to protect the underside of the foot and to ease locomotion.

Figure 2.11 Snail locomotion

Table 2.4 Animal support systems

Animal	Support system	Locomotion	Adaptation to environment
Earthworm	A hydrostatic skeleton. The fluid inside the closed body of the earthworm provides structure and support for shape.	Moves by using muscles to shorten and extend body, i.e. as in contract and expand.	Body adapted to life in damp soil. No protection from dehydration. Body shape suitable for environment.
Land snail	Shell and muscular foot. A hydrostatic skeleton similar to earthworm.	Movement through contracting and extending foot. Muscle movements on the underside of foot can be easily seen rippling.	Body adapted to life on land. Dehydration prevented by shell. Slime produced makes surface damp and movement easier (anti-friction).
Insect	Hard outer covering. Interconnecting plates with joints. (External skeleton or exoskeleton.)	Muscles attached to inside of skeleton, working on various parts.	Body adapted for life on land, air or water. Not likely to dehydrate as outer covering skeleton is water-proof. Insects are so successful because they can exist in many different habitats.
Fish	An internal skeleton which includes a backbone to which muscles are attached. The units are made of either cartilage or bone.	Muscles expand and contract, acting on the backbone and units making up tails and fins. Fins are from skin supported by skeletal fin rays.	Body is adapted for life in water. Respiration requires the extraction of oxygen from the water by the gills. Carbon dioxide is returned via the same route. Eggs, which have no protection from dehydration, are adapted to life in the water. Fish live in fresh and salt water. They exist in many different shapes and sizes, depending on their feeding habits and which level of the water they inhabit.
Frog	Internal bony skeleton with backbone. Four limbs with web of skin between toes of hind feet.	Powerful hind legs with well developed muscles for jumping and swimming.	Frogs spend most of their life on land, returning to the ponds to lay eggs. The skin is moist to allow some respiration, so frogs need to live in damp surroundings. Frogs have lungs for breathing air. Hind limbs have webbed toes for swimming. Eggs have no protection from dehydration, so are laid in water and develop there.
Bird	Inner bony skeleton with a backbone. Bones have air spaces within them to provide strong but light structures for flight. The part of the backbone in the neck is flexible, but the rest is fused to form a rigid frame for muscles used in flight to work against. Four limbs with fore limbs modified to form wings.	Walking or flying. Powerful muscles developed in chest area to maintain flight by acting against rigid backbone. Birds are capable of flapping and soaring.	The internal bony skeleton is highly adapted to flight, as is the very effective respiratory system. There is no dependence on water for life cycle, as the egg is protected with a shell and adapted to life on dry land. The outer part of the skin is adapted to produce feathers.
Mammal	A backbone or vertebral column to which bones, ligaments and muscles are attached.	Mammals can carry out a variety of forms of locomotion, flight, as in bats, swimming, as in dolphins and whales, and various types of movement on land.	This is a very successful group of animals, which can be found in many different environments and can sustain life in air, water or on land.

Plant support systems

In plants, shape and form is maintained mainly through the presence of specialised thickened cells within the stem which give strength to the plant structure (Figure 2.12). The strength is usually combined with flexibility so plants can bend without breaking. As in the case of the earthworm or snail, much of the support system for plants is dependent upon water pressure. When plants are deprived of water, they wilt as water is lost from the cells which will eventually collapse.

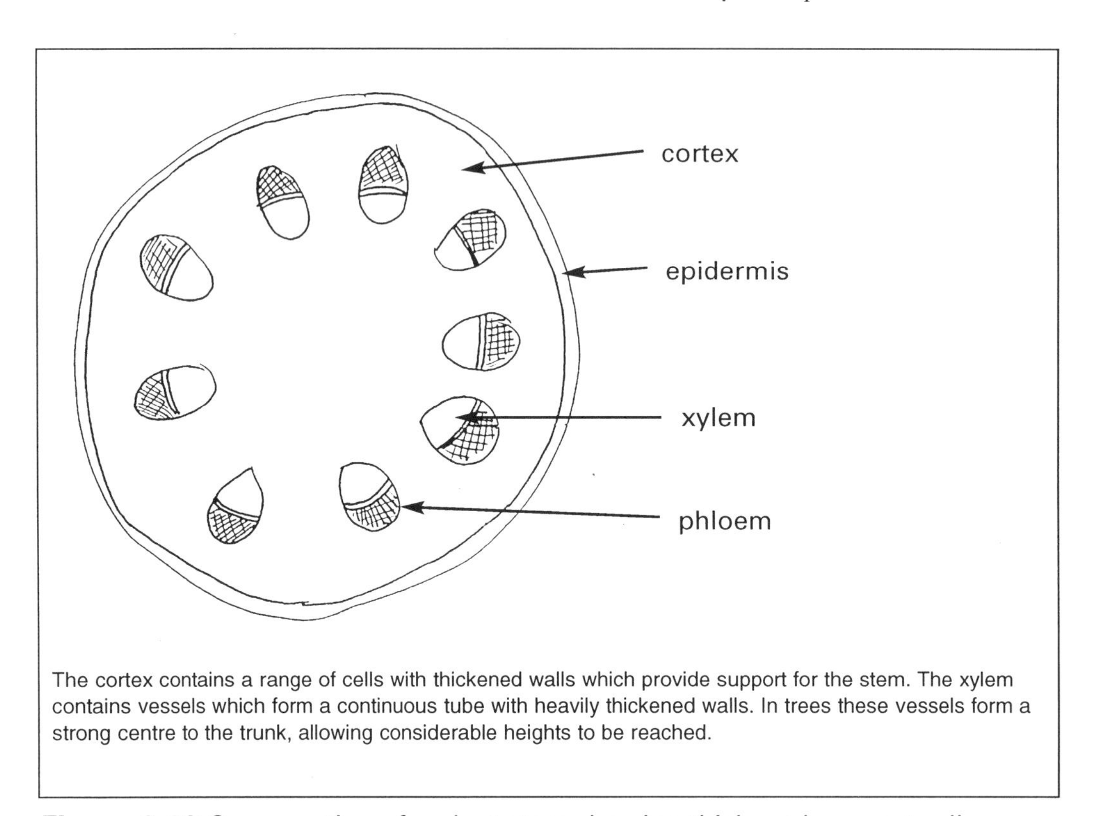

The cortex contains a range of cells with thickened walls which provide support for the stem. The xylem contains vessels which form a continuous tube with heavily thickened walls. In trees these vessels form a strong centre to the trunk, allowing considerable heights to be reached.

Figure 2.12 Cross-section of a plant stem showing thickened support cells

Excretion

Certain end products of the many chemical processes which take place within plant and animal cells would act as poisons if allowed to build up. For example, in respiration both plants and animals produce the waste products carbon dioxide and water. Carbon dioxide is absorbed into the blood and carried to the lungs in mammals or to the gills or skin of other animals, where it is returned to the surrounding air or water. In higher plants, carbon dioxide is lost through the leaves or used up in photosynthesis. The other major breakdown product which may build up within the bodies of animals is ammonia, produced by the breakdown of proteins. The form in which this is excreted depends on the group of animals.

Reproduction

Reproduction is a process whereby the species is continued, even though individuals involved will eventually die. Within both the plant and animal world there is a huge variety of life cycles and developmental patterns, but all species reproduce. Fungi, such as mushrooms, can reproduce asexually by spores, as can mosses and ferns. Asexual reproduction is usual in the lower orders of plants and animals. Flowering plants and most animals reproduce sexually to provide the next generation.

In all plants and animals where sexual reproduction takes place, specialised cells called gametes are involved. The male sex cell always fuses with the female sex cell to produce the zygote (e.g. a fertilised egg). In flowering plants, fertilisation is proceeded by pollination (Fig 2.13).

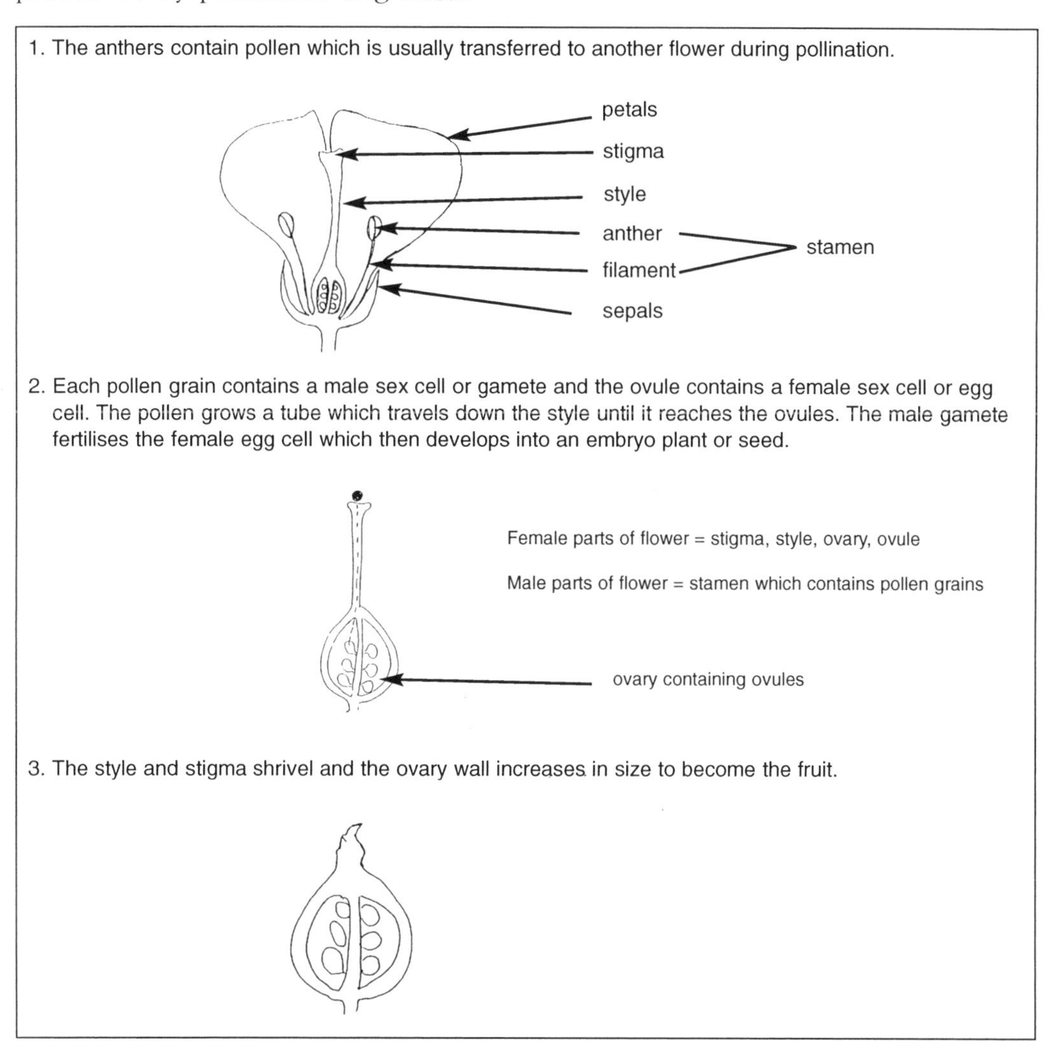

Figure 2.13 Parts of a flowering plant and sequence of events during pollination and fertilisation

By undergoing cell division and growth, the zygote develops into an embryo or developing organism. The embryo undergoes a varying length of development either externally (as in the frog) or internally (within the mother in many mammals), to become a free-living or dependent young of the same species.

Growth

All living things grow by cell division followed by an increase in cell size. In general, growth does not take place uniformly. In mammals, for example, the main growing regions are at the end of bones and in layers of cells in the skin. In flowering plants, growing regions are found mainly in root tips and buds.

In the early stages of development, all cells are able to divide, as in the germinating seed. In later developmental stages cell division is found in specialised areas of the plant or animal body.

Sensitivity

Sensitivity is the ability to respond to a change in either the external or internal environment and is essential for ensuring survival. When the stems and the leaves of a plant bend towards the light or the root grows down into the ground in response to gravity, the plant increases its chances of survival by maintaining an optimum environment. Animals respond to a wide range of stimuli including light, sound, and temperature.

Changes can also occur within the body of an animal and require some kind of reaction, such as muscle contraction. Survival is of the utmost importance, whether it is a response to a predator or simply to external conditions which might be damaging.

A study of a few familiar animals shows how they all share the ability to be sensitive to various conditions and also how their responses are particularly modified for the environment in which they live.

Children's understanding of life processes

Looking at children's ideas or concepts of living things is a well-established area of enquiry. For example, Piaget made an extended study of children's ideas of living and non-living and how things happen. Piaget, working with children who would now be considered to be at Key Stage 1, found that they believed that clouds moved by their own power across the sky and had some of the attributes of living things. This highlighted the fact that young children can move easily between the real world and that of imagination.

Many recent findings show that children do not employ a scientific approach to living things but use some life processes in isolation to determine whether something is living or not. Some examples are shown in Table 2.5.

It is very important to explore children's existing ideas before undertaking any work on life processes and living things. We may talk about plants as a major group, but it has been shown that children consider plants to be flowers found in

Table 2.5 Children's ideas of living things

Object	*Category*	*Reason*
Fire	Living	Moves, makes a noise, gives out heat
Bicycle	Living	We keep it in our hall; it lives in a house
Clock	Living	It makes a noise; it moves

gardens – otherwise they are weeds. Children also have a different definition for animals. They tend to equate animals with mammals – other members of the group often being referred to as 'creatures'.

A very vivid example of a young child's understanding of living and non-living could be seen when a group of children watched a tin which moved backwards and forwards (because of an internal mechanism of weights and a rubber band). One child decided that a hedgehog was inside the tin, making it move. Poor hedgehog.

It is worth while doing your own investigation into children's understanding of living and non-living. Identifying children's ideas provides a starting point from which we may develop their conceptual understanding through investigation, discussion, and bringing ideas out into the open.

Teaching points in life processes

When looking at the area of life processes and living things, it is important to identify relevant starting points and to develop a sequential path through the wide range of concepts underlying this area. Research has shown that children approach all areas of experience with their own ideas about particular concepts which may differ considerably from standard scientific concepts. Allowing children the opportunity to express their ideas will give us clear indications for starting points and ensure a match between children's understanding and the learning experience we provide.

Where do we start

The major concept underlying any work to be undertaken on life processes and living things is 'What do we mean by living?' This presents a wide range of exploratory work which can be readily undertaken in the classroom looking at ideas behind:

- living;
- once living;
- never having lived.

Thinking about these categories may help to clarify our own ideas about life processes and living things. As adults, we are usually able to distinguish between things that are alive, were once alive or have never been alive. Misconceptions can occur, e.g. seeds have been known to be classified as dead but will be alive in the future, so occupying a position of being living and non-living at the same time. Their appearance when dry does tend to encourage this response, but this impossible position cannot really exist. Seeds belong in the group of living things and carry out essential life processes.

Children, even at the end of Key Stage 2, do not appear to have a grasp of the full range of life processes, but this is a very complex area which needs to be approached in a progressively more detailed manner through first-hand experience.

Collections

Providing suitable collections of living and non-living material for children to explore and observe carefully in the classroom can usually provide a starting point, not just for discussion but also for many valuable investigations. Collections of material may be chosen by the class teacher with a particular aim in mind. However, collections which children have added to are more likely to gain their interest (see safety note at the end of the chapter). A typical collection might include:

- fruits and seeds
- leaves
- metal objects such as a spoon
- pine cones
- a flower
- a snail
- piece of wood
- bark
- a stone
- a fossil
- an apple

The collection should provide opportunities for children to bring their ideas out into the open for discussion. Providing magnifiers will help children to observe more closely.

A walk in the school grounds can also present children with the opportunity to identify living and non-living material and to observe materials within their natural surroundings (see safety note at end of the chapter). Any collection will present perplexing examples, for example, an apple which may be hard to classify but provides the opportunity to consider the idea of *once living; decay* and also *what about the seeds?*

Using open questioning can provide opportunities to develop aspects of the process skills of Science 1 (Sc1) together with children's understanding of the concept of living and non-living. For example:

- Where do you think this came from?
- What do you think this is made of?
- What do you think will happen if . . . we plant it/we put it in the soil?

For Key Stages 1 and 2 we can start with a fairly similar collection and then move children forward into progressively more detailed observation to include weighing objects and measuring their length or area. As we move through the

Table 2.6 Using collections to develop children's concepts

Flower	Structure – form and function Naming parts Comparison with other flowers Adaptation Life cycle
Seeds	Looking at structure of seeds Germination Growing seeds – conditions for life and growth Looking at different stages of development – life cycle Comparing and contrasting seeds – sorting and classification Seed dispersal
Leaves	Variety of shape Variety of colour Similarities and differences – sorting, classification Function

primary age range, children's questioning should become more detailed and complex, as should the tentative explanations (hypotheses) which they give when trying to provide answers.

A collection of living and non-living material can provide starting points for increasingly complex investigations, moving towards the identification and use of specific variables within experimental design (Table 2.6).

Children's questions

Children's questions may be used as valuable starting points for the investigation of living material, e.g. in relation to a collection of familiar animals such as earthworms, snails or woodlice.

- How does it move?
- What does it like to eat?
- Does it have a home?
- Is that trail to stop it getting lost?
- Are those eyes on the end of the feelers?

Children will provide many questions of this type, and if we are prepared to ask 'How do you think we can find out?' we are at the beginning of investigative work which will hold the children's interest.

THE DIVERSITY OF LIFE – CLASSIFICATION

Understanding classification

The living world is made up of millions of different organisms of varying form, structure and complexity. There are many different strategies for grouping or classifying living things, some more effective than others. If we simply use the external features of an organism or where it lives, as an indicator for making groups, we would end up with very different living things being grouped together. A very simple example of this would be trying to classify animals on the way in which they move. The examples in Table 2.7 are all able to swim but belong in very different groups.

Table 2.7 Swimmers

Animal	*Group*
Stickleback	Fish
Duck	Bird
Turtle	Reptile
Seal	Mammal
Whale	Mammal
Human	Mammal

A biologist uses a system of classification which takes both external and internal features into account. An effective classification system goes further than putting organisms into particular categories; it should also give an indication of the evolutionary pathway of particular groups from a common ancestor.

Although we tend to group life forms into plants or animals, there are now five accepted major groups or kingdoms as illustrated in Table 2.8.

Each kingdom can be divided further into small groups called phyla; members of a phylum have major features in common. The animal kingdom is divided into 25 different phyla. Each phylum is divided further into classes, and within each class are groups called orders (see Figure 2.14).

An **order** is broken down still further into groups called **genera**;. animals within a particular genus share many features in common. For example, the order Carnivora contains the genus *Mustelus*, which includes stoats, weasels, and polecats – a group which obviously has many similar features.

The smallest natural group of organisms which make up the sub-divisions of a genus is a **species**. Apart from small variations, members of a species are almost identical. One of the main features which determines whether organisms belong to the same species is whether they can successfully breed together and produce fertile offspring.

In order to have a classification system that is recognised throughout the world, there is a binomial system. This uses two words in Latin form: the first name gives us the genus and the second name gives us the species, rather like having a first

Table 2.8 Kingdoms of living things

Kingdom	*Characteristics*
Animalia	Animals are multicellular organisms. Most animals take in solid food as their form of nutrition (heterotrophic).
Plantae	Plants are made up of many cells. They have chloroplasts containing the photosynthetic pigment chlorophyll. They make their own food (autotrophic), channelling the energy from sunlight to combine carbon dioxide and water to produce simple carbohydrates.
Fungi	These are multicellular organisms which have a very particular structure of microscopic threads. They may be parasitic (living on other organisms) or saprophytic (feeding or living on dead organic materials); includes mushrooms, toadstools, bracket fungus, mould on stale bread or cheese and the yeast which we use for brewing and baking.
Monera	Usually unicellular and relatively simple in structure. This group includes bacteria and blue-green algae. Can cause disease or ruin food, but most are harmless or can be very useful, in decay they play a vital role in recycling of nitrates in soils. Viruses also belong to this group, although they are debatable living organisms as they lack some of the functions of living things. They reproduce only inside the cells of other living organisms, can withstand extreme temperatures and can exist in the form of crystals.
Protoctista	Relatively simple body structure, generally single-celled but some multicellular. Nutrition can be heterotrophic, autotrophic containing chlorophyll or parasitic, e.g. *Plasmodium* which causes malaria.

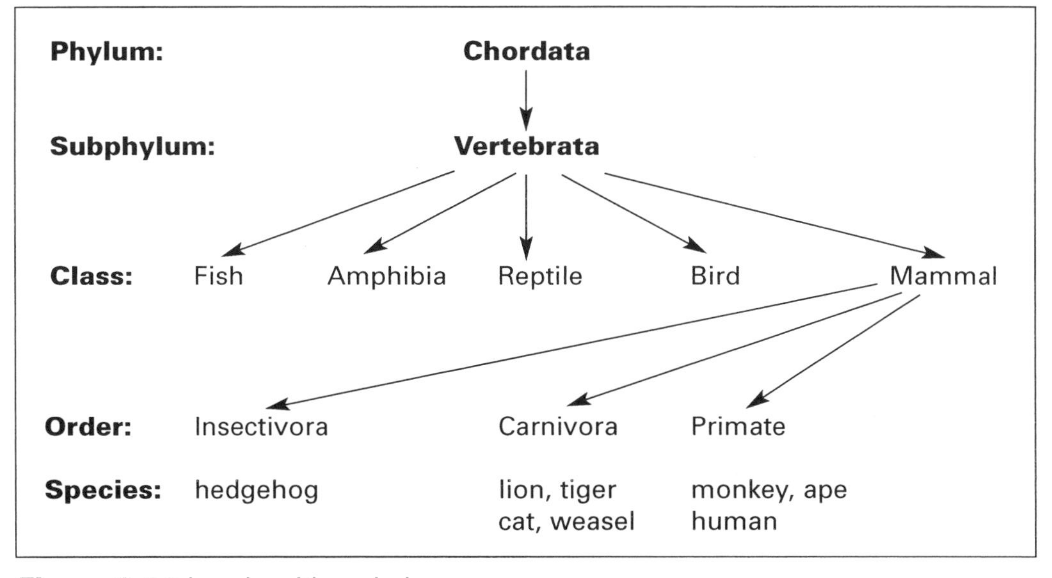

Figure 2.14 Levels with a phylum

name and a surname. Taking our previous example of the order Carnivora:

- Genus: *Mustela;*
- Species: *Mustela erminea* (stoat), *Mustela nivalis* (weasel).

Table 2.9 shows how this classification system works for the cabbage white butterfly.

Table 2.9 Classification of cabbage white butterfly, *Pieris brassicae*

Phylum	Arthropod	External skeleton; segmented body with jointed legs; marine, freshwater, terrestrial
Class	Insect	Three pairs of jointed legs; compound eyes; usually two pairs of wings
Order	Lepidoptera	Small to large insects with entire covering of powdery scales; metamorphosis complete
Family	Pieridae	
Genus	*Pieris*	
Species	*brassicae*	

Using keys for identification

An identification key can be made up using almost any characteristic of an organism. However, its success can be judged on how effective it is in providing a correct identification. Using a key for identification purposes always requires careful observations of the particular plant or animal and making decisions based on these.

There are two main ways in which keys may be constructed. The simpler of the two is the decision tree where questions are posed which require a yes or no answer. For example, let us look at a fairly common collection of animals:

- cabbage white butterfly
- worm
- snail
- woodlouse
- ant.

A decision tree about these animals can be laid out as in Figure 2.15.

An alternative form of key is the dichotomous or statement key which has two branches so that you are confronted with two possibilities at each stage. These possibilities are expressed as statements or clues, one of which must be chosen in order to lead on to the next pair of statements or clues. An example is given in Figure 2.16. The statement key is used by biologists during field studies and is very much more accurate than a decision tree, in that it provides much greater detail.

When working with children, their first introduction to classification should be

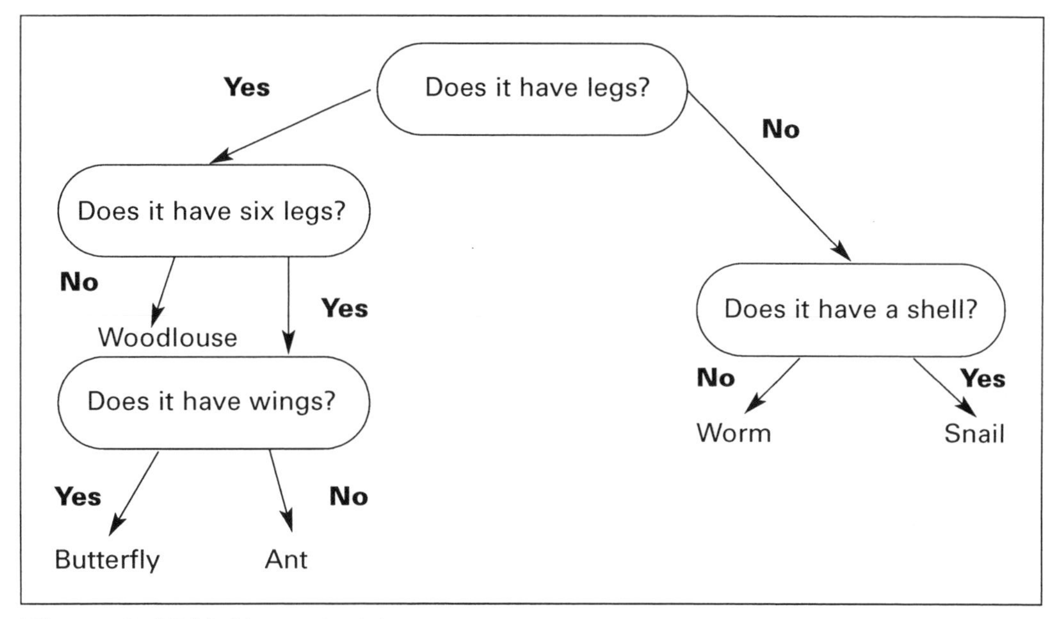

Figure 2.15 Making a decision tree

through the simpler decision tree approach which deals with a limited range of observable features.

1	Jointed legs present	EITHER go to clue number	**2**
	No legs	OR go to clue number	**15**
2	3 pairs of legs	(INSECTS)	**4**
	More than 3 pairs of legs		**3**
3	4 pairs of legs	(ARACHNIDS)	**9**
	More than 4 pairs of legs		**11**
4	Wings present	Plant bug (Order Hemiptera)	
	Wings absent		**5**

Figure 2.16 Part of dichotomous or statement key (adapted from Phillips, W. D. and Chilton, T. J. (1993) *A Level Biology.* Oxford: Oxford University Press)

Children's understanding of classification

Children do not classify or group in a particularly systematic way when left to their own devices. For example, when carrying out an investigation on 'minibeasts', they will not necessarily see the link between these and other animals. Young children may group in terms of colour, shape, ways of moving, etc. Animals which are found living close together may be defined as a group. Children will use various strategies appropriate to their level of development and understanding.

Teaching points in classification at Key Stage 1

Assuming that children have carried out investigations or had experience of the concept of living and non-living, it is valuable to move on to provide them with further collections to sort, group and classify and to encourage them to recognise similarities and differences. Any work on grouping will encourage the beginnings of classification, whether the material is living or non-living. This links in very well with the early development of children's understanding of materials.

Collections

We can start with:

- Photos or drawings of living organisms with which they are familiar.
- Various objects from the living world: seeds, leaves, fruits and 'vegetables'.
- Living material encountered within any study of school grounds or nature area.

Discussion
A starting point for all scientific work with children is to allow them to voice their own ideas. Working with a carefully chosen collection, the teacher can provide opportunities for children to discuss their existing ideas and, through careful questioning techniques, move these ideas forward.

Encouraging observation
When children are provided with any collection of material, the amount of visual information they receive can be overwhelming. It is important to help children focus on relevant detail by providing guidance through the use of open questioning, for example:

- Can you see . . . ?
- How do you think it moves?
- Where do you think this comes from?
- How many legs can you see?

Children's questions
Children can be encouraged to look again at material or to move forward in their observation by using their own questions as a source of enquiry. When presented with a simple collection of minibeasts (e.g. snails) children's questions tend to be of the following type:

- How does it move?
- How does it see?
- Are the little ones 'babies'?
- What does it eat?

Incorporating children's questions such as these into wall displays or home-made books, encourages them to present tentative explanations (hypotheses) and to realise the need to observe again and more closely, with perhaps the aid of a magnifier.

Information gathered during such observations can lead to noting similarities and differences, which is the basis of any form of classification. As children develop these skills, they become able to provide simple statements which then form the basis of a decision tree. The essential skill is that of providing questions which may be used to separate members of any one group. This skill becomes increasingly better developed as children progress through Key Stage 2.

The use of games

Games can often be used to promote the development of specific skills within areas of science. For classification, children can be encouraged to make a card of a number of simple statements for an animal or plant guessing game. A simple example could be:

- I fly
- I have feathers
- I lay eggs
- What am I?

Children can provide cards such as these to go with a group of minibeasts under investigation. Games which particularly encourage the development of the principles behind classification are the 'Who?' or 'What am I?' games. These are particularly suited to children at Key Stage 2, as there is a need to develop a fairly complex, logical approach. Children decide to be an animal or a famous person and, in trying to find the answer, other children must use questions of the type which form the basis of decision trees, i.e. with yes or no answers only. An example is given in Figure 2.17.

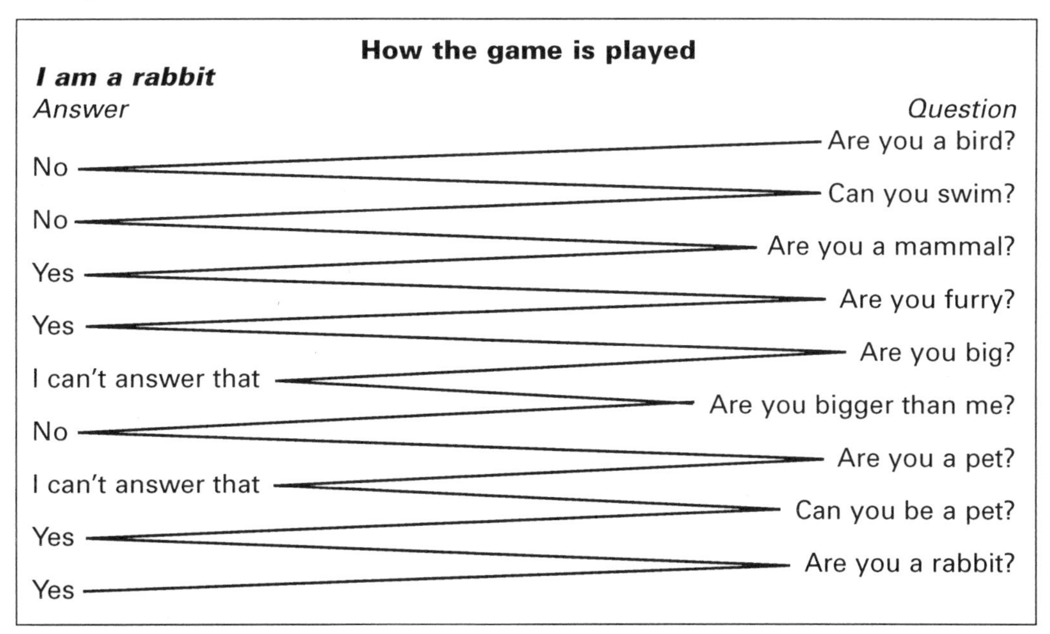

Figure 2.17 Classification game

ADAPTATION

Understanding adaptation

When we look at the living world around us, we cannot help but notice the enormous variety of shape and form. In order to make some sense of the way in which this interdependent system functions, it is useful to view it through the interconnecting principles of **form**, **function**, **adaptation**, and **interdependence**.

All living organisms show adaptation for a particular way of life within the environment. Adaptation can be looked at in terms of body shape and form, and the way in which all, or parts, of the body of a living organism are modified for a particular purpose. These modifications relate to how an organism provides for its needs within the environment. We term this suitedness 'adaptation'. A fuller picture of adaptation also includes the idea that adaptive characteristics are inherited and not acquired during the lifetime of an individual organism – in fact they are found within all members of a particular species. When these inherited characteristics allow the species to function successfully within the environment, the species will survive and continue. Should the environment undergo changes, a particular species may no longer be suited to the new conditions and may eventually die out and become extinct.

The concept of adaptation can be explored through reference to all living organisms. Table 2.4 provides some well-known examples. Plants provide an excellent area of study which may help to develop the concept of adaptation. If we look at plants, we are immediately provided with a wealth of material for children to explore (Figure 2.18).

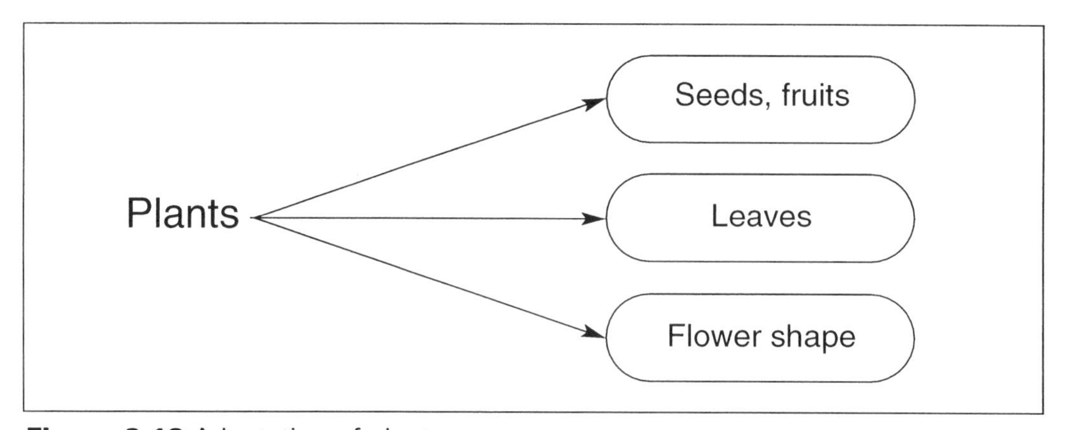

Figure 2.18 Adaptation of plants

Fruits and seeds

It is essential for all living things that the species survives, and plants are no exception. The seed is always the result of sexual reproduction and is the source of the next generation. Different fruits and seeds vary enormously, for example, poppy seeds are numerous and very small when compared with the large coconut.

Table 2.10 Seed dispersal mechanisms

Dispersal of seed	Adaptation
Wind dispersal	
	Poppy Capsule jerks backwards and forwards ejecting hundreds of small seeds out through a number of small openings.
	Sycamore Increase in surface area to reduce rate of fall so that seeds can be caught up by air movements.
	Dandelion A feathery parachute mechanism.
Water dispersal	
	Water lily Has a seed with spongy extensions which allows it to float in water.
	Coconut Has a fibrous outer coat which allows it to float in water.
Animal dispersal	
	Berries and **plums** Eaten by birds.
	Cleavers and **burdock** Hooked fruit cling to fur of animals.
	Nuts and **acorns** Carried away and buried by rodents.
	Gorse Oily seeds transported by ants.
Explosive mechanisms	
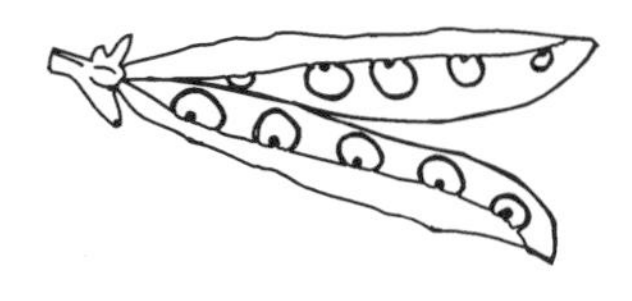	**Pea** Pod dries out, twisting as it does so.

Seeds can be found with wings or parachutes attached or embedded in the flesh of fruit.

A study of adaptation of fruits and seeds in relation to dispersal shows a range of mechanisms which help to ensure survival. Young plants, particularly the offspring of large trees, need to develop in an area where they are not competing with the parent plant for light and space. Distances travelled by seeds vary, but mechanisms involving water or animal dispersal often result in young plants developing in and colonising areas at a considerable distance from the parent plant (Table 2.10).

Leaves

Leaves occur in many shapes and sizes and a study of them will highlight the way they may be adapted to function successfully in a range of areas. Climbing plants have leaves modified to form tendrils which cling to other sturdier plants, allowing the climbers to obtain greater height and support; this also provides maximum exposure of the leaves to sunlight.

In very dry arid areas, leaves may be modified for water economy, as the leaf is normally the source of water loss. In many cacti, the leaf surface is reduced to form spines. Leaves may also be modified to form prickly spines as protection against animal feeders, e.g. gorse.

Flowers

It is immediately obvious that flowers occur in many different shapes, sizes and colours. Although we may admire the beauty of a particular flower, its most important function is reproduction, i.e. survival. Flowers must be pollinated, usually by other plants of the same species, which requires pollen to be transferred from one flower to another. There are various methods of pollination, but each flower is highly adapted to ensure that this is carried out effectively; examples are shown in Figure 2.19.

The range of adaptations to be found in plants is very extensive and can only be covered in part here. Video clips can be used to show pollination and dispersal mechanisms in action in exotic places as well as more local ones. The BBC video *The Secret Life of Plants* beautifully illustrates a range of these mechanisms in great detail.

A study of one particular major group within the animal kingdom can be used to illustrate the general principle of adaptation. The phylum Arthropoda (Figure 2.20) provides a fascinating study of variety, form and function, adaptation and diversity. The name Arthropod means jointed limbs and this group includes lobsters, crabs, flies, butterflies, bees, pond skaters, water boatmen and spiders. The list is extensive, but it does include many of the animals children encounter every day, particularly when they are pond dipping. All members of the phylum have jointed legs and hard external skeletons which can be waterproof.

The background information in Table 2.11 gives some idea of the range of adaptations which have taken place to allow individual members to exist successfully in a wide variety of habitats.

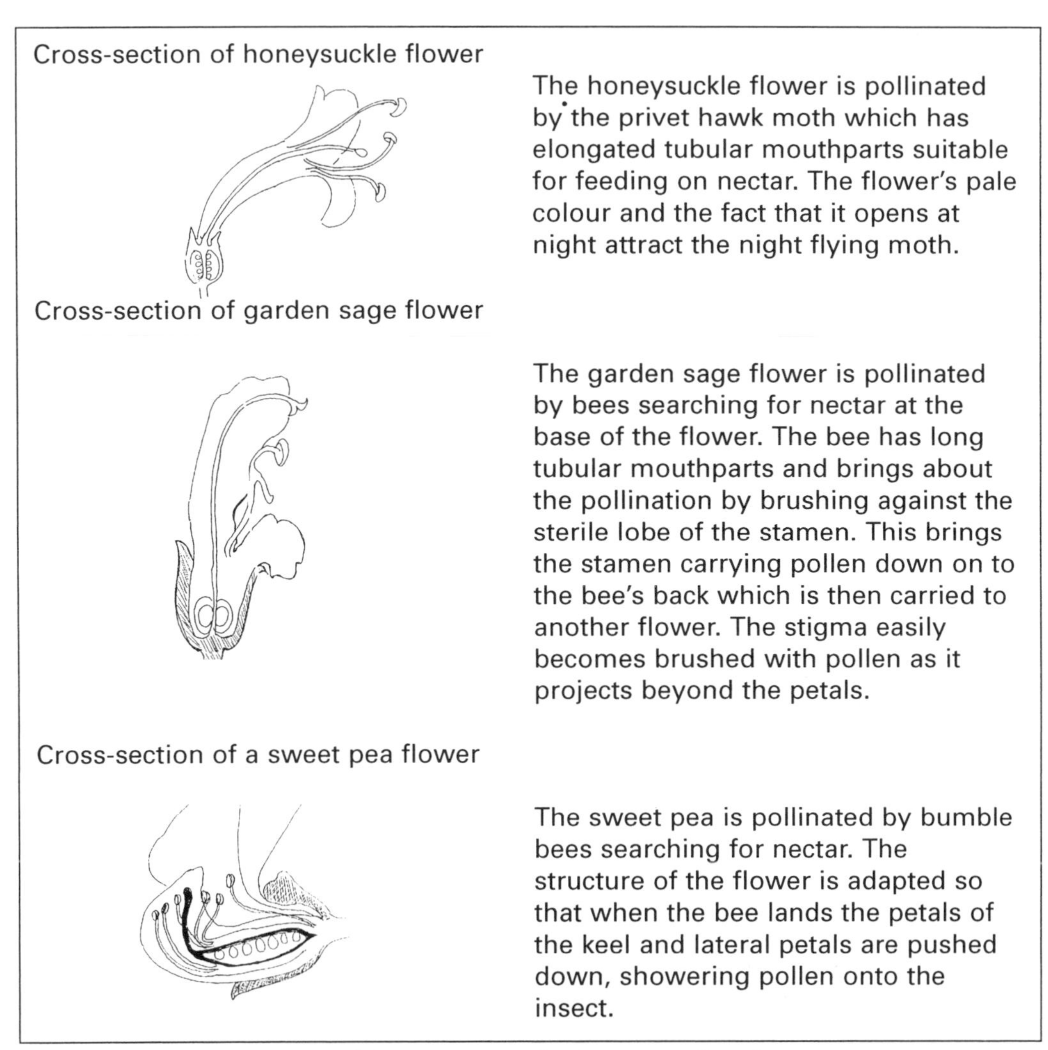

Figure 2.19 Adaptation of flower shape for insect pollination

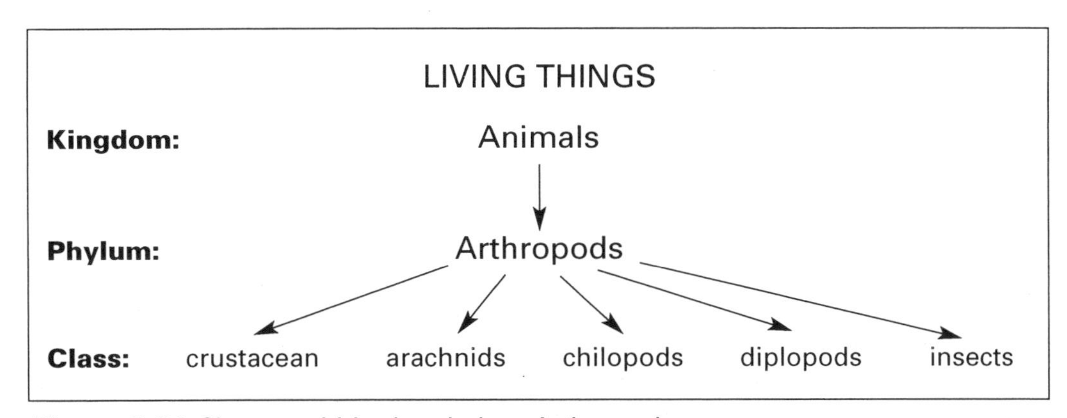

Figure 2.20 Classes within the phylum Arthropoda

Table 2.11 Phylum Arthropoda

Class	*Characteristics*	*Examples*
Crustaceans	Usually found in fresh or salt water. Very few are terrestrial, as outer skeleton is not waterproof or suitable for dry conditions. The woodlouse is an exception as it lives on land but can only survive in dry conditions.	Lobster, crab, waterflea, woodlouse
Arachnids	These are all terrestrial and can exist in dry conditions.	Spider, scorpion, mites
Chilopods	A terrestrial group which are herbivores, eating leaf litter and soil, and moving fairly slowly.	Centipedes
Diplopods	These are adapted for more rapid movement than centipedes and are predators.	Millipedes
Insects	Predominately terrestrial with the exception of the midge, which lives in the sea. These are an extremely successful group of highly adapted animal which show enormous variety of shape and form.	Housefly, bee, wasp, pond skater, water boatman, ladybird, greenfly, beetles

Insects

Insects are an example of a highly successful group which exhibit a very wide range of body form, adapted for specific lifestyles. They are found in a variety of habitats throughout the world – in the air, on land, and on water. Within this class, a wide variety of size, shape, colour, mouthparts, feeding habits, legs and methods of locomotion can be found. Looking at shape, Table 2.12 highlights a few well-known examples.

Within the group, shape and form can be adapted not only for various forms of locomotion but also for concealment and mimicry. Each insect's shape and form suits its purpose and place within the natural world.

The range of feeding habits found in insects is also extensive. This is reflected in the highly specific adaptations of mouthparts. Figure 2.21 gives only a few examples, but looking at the mouthparts of the bee and butterfly shows not only how these are adapted for a very specific method of feeding, but also highlights the independence between these and the adaptations found in insect-pollinated plants.

Table 2.12 Insect adaptations

	Example	*Shape and locomotion*
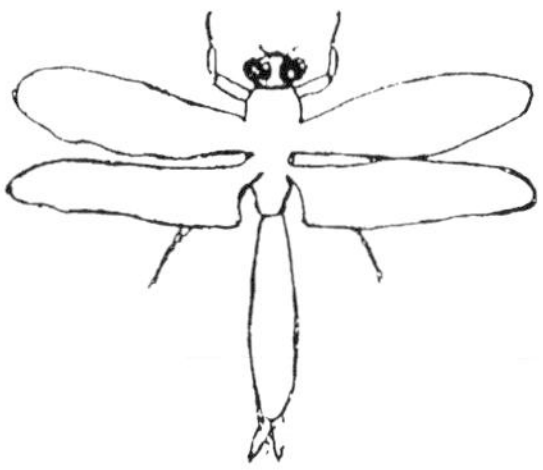	**Mayfly** and **dragonfly**	Have an attenuated shape with wings and are adapted for flight.
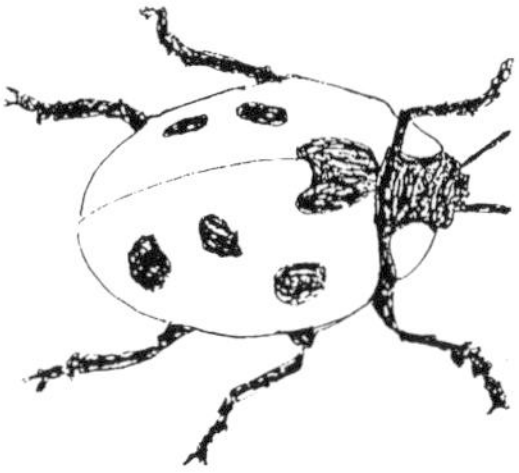	**Ladybird**	Has a bulbous, rounded body, suitable for crawling and limited flying.
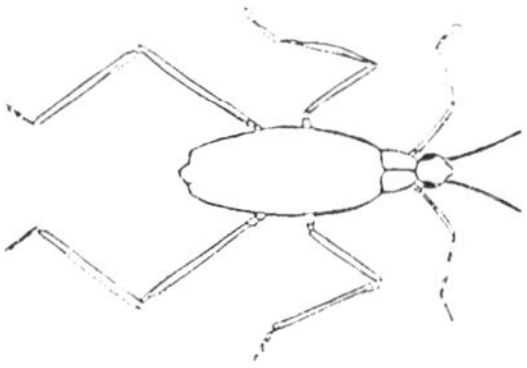	**Pond skater**	Has a narrow light body, with fine short hairs which makes the body unwettable and allows the skater to move easily on the surface film of water.
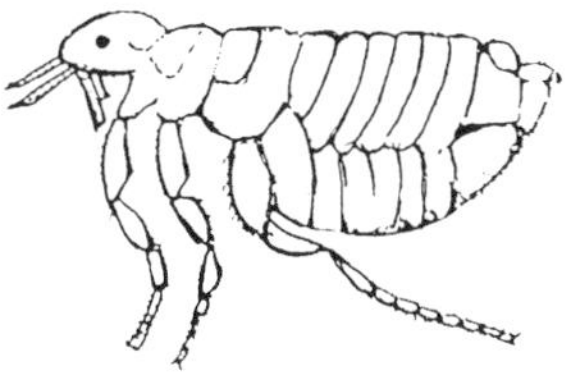	**Flea**	Has a flattened body to ensure easy movement among animal fur.
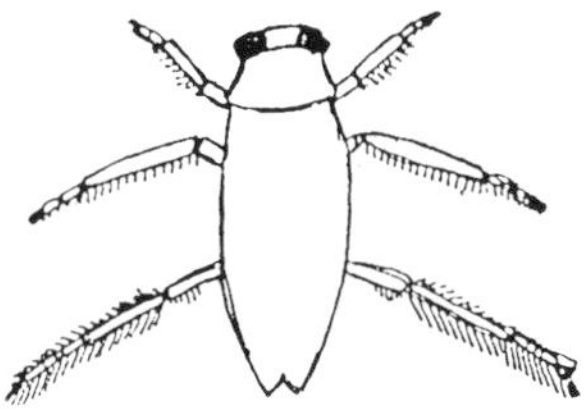	**Water boatman**	Has long hind limbs with fringes of bristles. These act like oars and row the insect along, lying on its back.

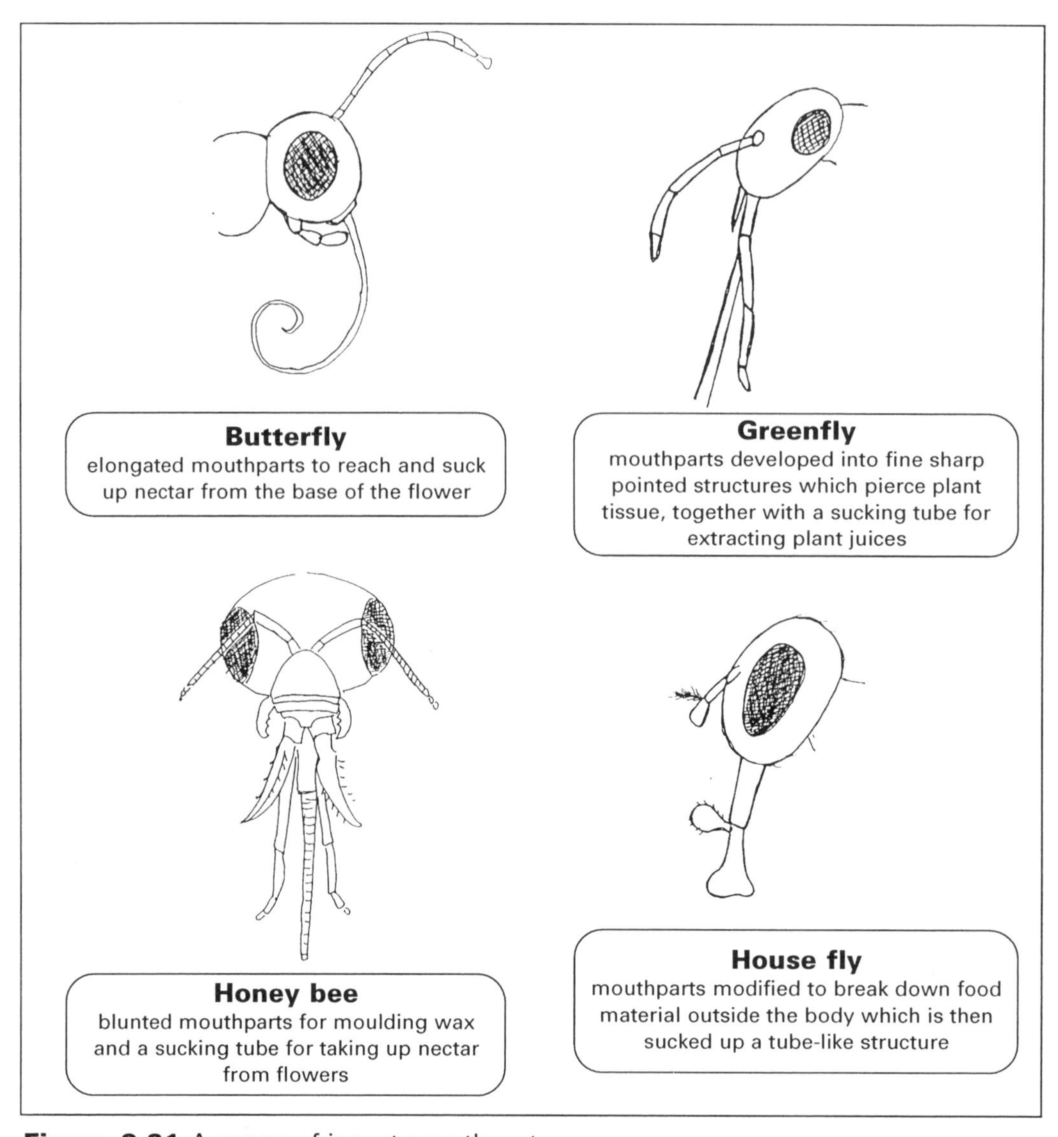

Figure 2.21 A range of insect mouthparts

Children's understanding of the concept of adaptation

The concept of adaptation is not an easy one to grasp. At Key Stages 1 and 2, children may see adaptive features as characteristics which just exist and relate to a particular organism rather than the species. In order to have some appreciation of the biological sense of adaptation, children need to have developed not only the concept of life cycle and like producing like, but also the overriding need for survival of the species. Without this understanding it is possible to develop the false notion of particular adaptations developing during the lifetime of a particular organism, rather than being an inherited characteristic. Children also need to be aware of environmental factors and the interrelationship between these and living organisms.

Teaching about adaptation

Adaptation is a complex concept which requires an understanding of structure and function, life cycle, inheritance, species and environment and may not be fully understood even at secondary level. However, if we identify the building blocks which underpin this concept we can provide valuable areas of study for children at Key Stages 1 and 2, which will lay the foundations for further understanding in later years.

Since adaptation can be observed in all living organisms, in terms of being suited to their environment, it is possible to provide children with the opportunity to explore aspects of this concept even without access to large, open spaces. Even in inner city schools, playgrounds can usually be modified to become an area of investigation of living things.

The following examples can provide opportunities for children to investigate aspects of adaptation and fitness for purpose:

- a bush: attracts insects and birds;
- a tree: provides a habitat for insects and birds and examples of seed dispersal;
- stones: provide a range of habitats;
- logs: provide a range of habitats;
- flower beds: provide a range of adaptations for pollination and seed dispersal;
- seed box: provides opportunities to observe life cycles;
- bird area: provides opportunities to observe adaptations for feeding and flight.

Children can be overwhelmed by an excess of material and information, so that investigating specific examples provides a simple starting point for developing this concept. Birds present a very particular example of adaptation of body form in relation to flight. A mixture of observation, use of videos and personal research can provide children with increased understanding and a chance to explore structure and forms adapted for flight within the animal world.

When providing any planted area for children to investigate, it is important to relate the variety of plants used to the need for children to develop particular concepts. Observing plants in their natural surroundings provides material for investigation which could not be carried out in the classroom. Observation of planted areas provides opportunities to note not only detail of form and function, but also life cycles of flowering plants. Planting seeds outside encourages children to carry out observations over a period and to develop appropriate methods of recording. Seeds which fail to thrive can encourage discussion of the needs of plants. It may also provide opportunities to observe the feeding habits of birds.

The use of open questioning can encourage children not only to observe in more detail but also to note the conditions where living things are found (see section 'Living things in their environment' for greater detail). The following questions may encourage children to begin to consider how an animal is suited or adapted to its environment.

- Where did we find this?
- How does it move?

- Which parts of the body does it use for movement?
- How does it use them?
- Can it move in different ways?
- What does it eat?
- How does it obtain food?
- Does it bite, suck, chew, etc.?
- How is it suited to the way it lives?

Questions which cannot be answered by direct observation lead children on to use appropriate reference books. Children can then produce a profile of a particular animal based on their own observations together with other relevant information. Getting children to consider the places where plants and animals can be found allows for the exploration of the idea of multiple habitats, e.g. in the air, in trees, on leaves, under stones or logs, on the ground or under the ground.

A detailed study of any particular animal can help to develop the concept of adaptation. The main questions being explored are:

1. How is this animal suited to living where it does?
2. How does it protect itself?
3. What mechanisms ensure survival?

These could lead on to investigating camouflage in a variety of ways, providing opportunities for testing a range of mechanisms and their effectiveness.

When developing major concepts with children at Key Stages 1 and 2, it is important to provide not only first-hand experience but also opportunities for discussion which allows children to extend their existing ideas. For example, opportunities to think about and discuss which factors may affect the survival of living things can lead on to consideration of changes in the environment either through natural or human influences. The idea of predator-prey relations could also be considered.

Life cycles and adaptation

A study of life cycles raises several important issues for children to explore at Key Stages 1 and 2, some of which contribute to an understanding of adaptation and others which develop the idea of like producing like. The diversity to be found within the life cycles of living organisms is considerable. Each life cycle, however, is adapted to suit the survival needs of the individual species.

A useful starting point for looking at life cycles is from the viewpoint of 'ourselves'. Getting children to record, in a simple form, the stages of the human life cycle not only helps to develop the idea of sequencing, but also reinforces the idea of different needs at different stages.

Comparing our own life cycle with that of other living organisms can help to highlight a number of similarities and differences. The idea of the need for protection in the early stages of an animal's life cycle and how particular species show adaptations to this is an interesting area to explore.

The life cycle of the butterfly or frog are often studied in schools as examples of metamorphosis, i.e. a complete change of body form during the completion of one life cycle. Both provide excellent examples of adaptation, together with areas for exploration.

Exploring adaptive features within the life cycle of the cabbage white butterfly can raise the following questions for the different stages:

- Egg: Where is it found? Why do you think this is a suitable place?
- Caterpillar: Where is it found? How does it move and feed? How is it suited to its environment? Is it in danger of being eaten? Who by? Is there anything about it which helps it to survive?
- Pupa: Where is it found? How is it protected?
- Butterfly: Where is it found? How does it move and feed? How is it suited to its way of life?

The same sort of pattern of discovery can be applied to the life cycle of the frog. Here, answers to questions will provide the idea of many eggs being produced to ensure survival of the species. The fact that up to a thousand eggs may be laid by a single frog gives some indication of the risks to survival. When children compare the structure, form and life style of the tadpole with that of the adult frog, they will become aware of two completely different responses to adaptation.

A study of fruits and seeds can also help to promote children's understanding of adaptation. Children can explore the link between the variety of form and function with the overriding need for survival. The adaptations are numerous but a study of a limited number of fruits and seeds can lead to a wide range of investigations, for example exploring winged fruits such as sycamore.

A study of these fruits and their behaviour when dropped from a height gives a clear indication of the function of the wing. Children observing winged seeds notice that they move slowly to the ground in a spiral fashion, allowing the wind to disperse them. Moving on from these observations, children can investigate the distance travelled by seeds from the parent plant, i.e. the effectiveness of this adaptation. Observing the behaviour of winged seeds also provides opportunities for hypothesising, predicting, and fair testing, e.g. children can experiment with reducing the size of the wings or the effect of removing them completely.

Altering the weight of the seed (e.g. by adding Plasticine) provides a further series of valuable explorations. Parachute mechanisms, such as those found in the dandelion, can also be explored and their behaviour recorded. Questions such as 'How does this help the seed to survive?' can lead on to discussion and further investigation. This particular area links very well with movement through air and the concept of air resistance. Children can move on to explore parachutes and autogyros which use parallel mechanisms. Making links with other areas of exploration provides opportunities for reinforcement of the underlying concepts. The range of opportunities to explore adaptation in fruits and seeds is considerable and other useful examples are noted in Table 2.10.

Observation of flower form and structure in relation to pollination can provide several examples of very specific adaptations. Children can observe which insects visit which plants and note whether there appear to be preferences for colour or shape. Observations can lead to a detailed exploration of flower parts, together with naming of parts and identifying their functions. Children may also explore whether scented plants attract particular insects.

LIVING THINGS IN THEIR ENVIRONMENT

Understanding the relationship between living things and their environment

Ecology is the study of living organisms in relation to their total environment, to include living and non-living aspects. The value of studying living organisms within their natural environment is that it provides the opportunity to see how plants and animals are adapted to exist in very particular conditions. A much clearer understanding of form and function can be gained, together with aspects of how particular life cycles are adapted for a specific way of life. In the laboratory or classroom, without normal environmental conditions operating, animal behaviour may appear to be random rather than highly adapted. An example of this can be seen in the woodlouse, an animal normally to be found in damp, shady conditions as it lacks a total waterproof covering. Within a classroom, woodlice exposed to light, dry conditions will move rapidly and apparently randomly. This does not give a clear impression of their normal behaviour. The movement often seen in the classroom is in fact highly protective, as these movements increase the likelihood of finding damp dark conditions as fast as possible. The needs of living things is a central issue here and studying natural environments gives children the opportunity to move beyond an egocentric view of the living world, that is from their own particular needs to those of living things.

A number of specific terms are used when studying living organisms within the natural environment. **Ecology** is essentially the study of ecosystems, a study which encourages the development of concepts such as interdependency, variety and adaptation. **Ecosystem** includes all the living and non-living parts of the environment; e.g. air, water, soil, light, plants and animals; it may be large (e.g. a lake) or small (e.g. an oak tree). The central features are that an ecosystem is a dynamic system which maintains an equilibrium or balance. Changes in one part of an ecosystem result in changes in other aspects so a balance is maintained. When the balance between all the interdependent units within the ecosystem is disturbed, a balance will be maintained by changes in another aspect of the ecosystem. If changes which are extremely drastic take place and the system cannot compensate, enormous damage can occur within particular areas, with entire species being wiped out.

A good example of an ecosystem is that of a woodland. This contains trees, smaller plants, animals, fungi and micro-organisms together with soil made up of rock particles, leaf and animal litter. The woodland ecosystem also includes climatic factors such as rainfall, sunlight and physical factors such as acidity or alkalinity of the soil.

Environment

Environment means everything in the surroundings of a particular living organism. It provides everything needed to support the life of the plant or animal, as well as including aspects that may prove harmful. Figure 2.22 shows an example of the environment of a fish. The term environment does not include the complex interrelationships and interdependence within the concept of ecosystem. The environment is the combination of external factors affecting a particular organism.

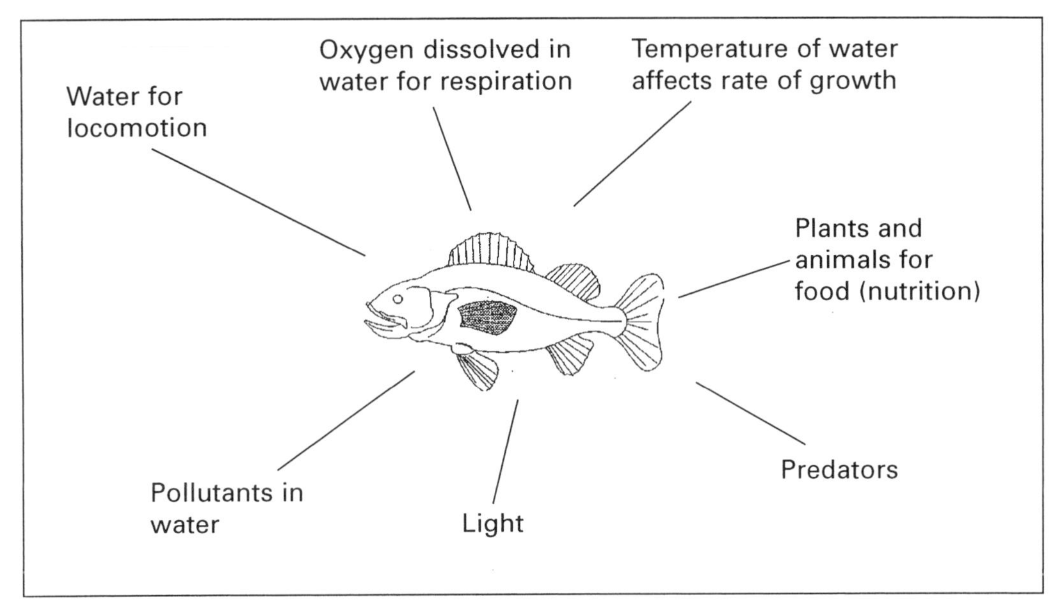

Figure 2.22 Environment factors

Habitat

Habitat is where the plant or animal lives, where it obtains food, shelter and where it reproduces. For example, the habitat of the greenfly is the stem or leaf of a rose bush (Figure 2.23). Unlike the term environment, habitat does not include climatic conditions (such as sun, wind and rain) or predators (such as the ladybird).

Population

This term always refers to a single species living together in a particular environment, e.g. we can refer to the population of sticklebacks in a pond.

Community

This term applies collectively to all the living organisms to be found within any ecosystem, e.g. a pond can be viewed as a community of plants and animals. A pond will provide a home and supports a way of life for many living organisms. A system of complex, interdependent relationships exists between the various species of plants and animals which are to be found there (Table 2.13). When taking a closer look at the living organisms to be found in a pond, it is possible to see that each occupies a particular habitat within the physical environment, with special adaptation for that way of life.

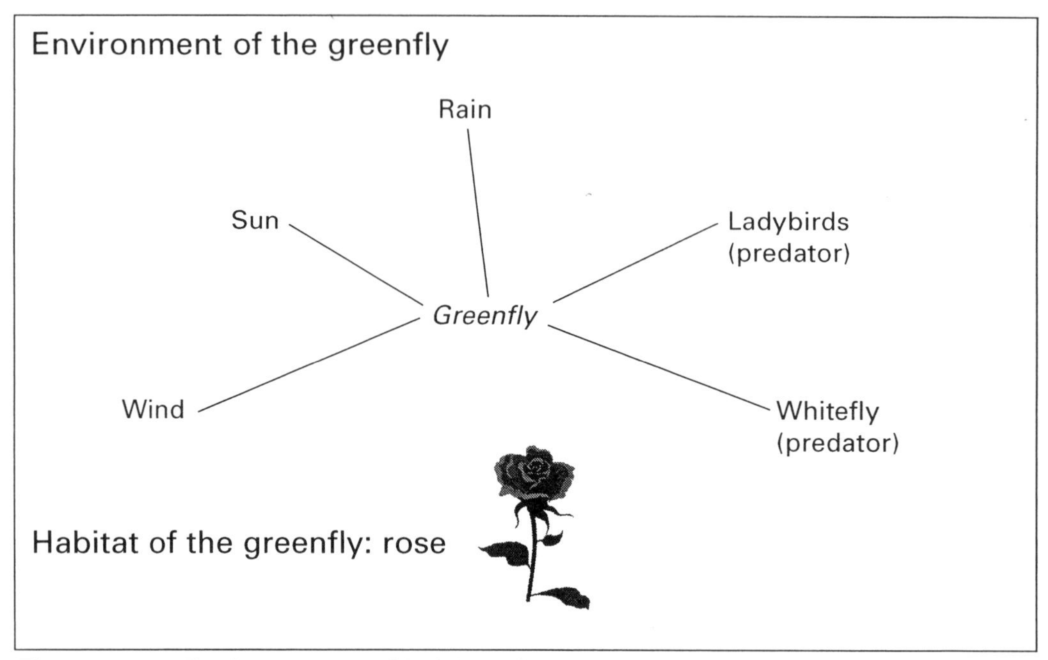

Figure 2.23 Environment and habitat of the greenfly

Table 2.13 A pond community

Levels	*Examples of animals found*
Surface film	Pond skater, springtail – *moving on the surface* Mosquito and phantom midge larvae – *feeding* Flatworms – *feeding and moving*
Top layer from 2–3 cm	*Cyclops*, plankton, water fleas – *feeding*
Middle layer	Water beetles, water boatmen, minnow, stickleback – *free swimming*
Bottom layer	Caddisfly larvae, water louse, freshwater shrimp – *feeding*
Mud	Freshwater mussel, pea shells – *feeding*

Children's understanding of the environment

In the early stages of Key Stage 1, children tend to view the world around them from their own point of view. Thus flowers are there to make gardens beautiful or to smell nice and fruits grow for us to eat. This egocentric view is part of a child's natural development. Teaching and learning about the environment during the primary years should enable children to move step by step towards an understanding of more complex issues. Young children generally see animals as individuals dependent on humans in some way. Children can develop enormous empathy for the needs of living organisms, but this tends to relate very much to their own needs.

Exploring the environment in a progressively more detailed manner allows children to move towards some understanding of our own position within a complex interdependent system. Towards the end of Key Stage 2 children are usually able to understand some aspects of the dependence between organisms in terms of feeding relationships, as in food chains, and the need for homes and shelter. The dependence on green plants as primary producers of food is a more complex issue which will need to be developed further during their secondary years. However, these two areas are well worth exploring, not only in terms of developing children's investigative skills, but also because these are the stepping stones towards building up greater conceptual understanding.

An understanding of the complexities of how climatic conditions affect plants and animals within an ecosystem is only just beginning to develop by the end of Key Stage 2. Simple examples may be explored but this area is more likely to be developed effectively during the secondary years.

Teaching about the environment

The study of living things in relation to their natural environment provides children with the opportunity to observe plants and animals carrying out life processes in their natural homes whether this be on land or in the water. A study of relatively few plants and animals in their natural habitat will encourage the development of a number of concepts.

A study of the world around us can be used to move children from the egocentric view of the world being there for us, towards an understanding of the way in which all living things have an interdependency, that each has its own needs and is adapted for a particular way of life. Study of the environment can increase children's understanding of a range of concepts as shown in Figure 2.24.

Assuming children have had some experience of the living organisms to be found within their local environment, it is possible to develop their understanding further by introducing the idea of 'homes' or 'where do I live'. At Key Stage 1, when asked to design a home for a 'minibeast', children initially think about their own needs for warmth and comfort, and want to provide beds of cotton wool and other soft material; fine for us, but perhaps not the best place for a snail or earthworm.

The most valuable aspect of children's understanding in the early stages of Key Stage 1 is their enormous empathy for 'small creatures' combined with a strong protective instinct. If this is channelled appropriately, care and respect for living organisms will be maintained throughout later years. Any work within the environment should enable children to understand the complexity and needs of

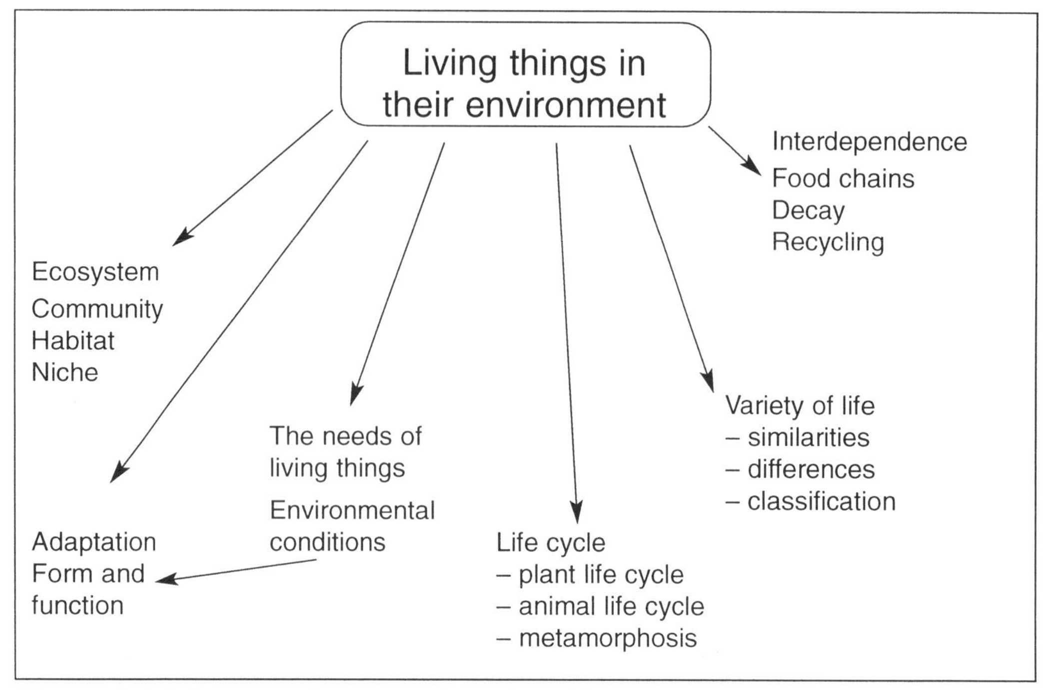

Figure 2.24 A range of concepts introduced by studying the environment

any living organism so that they are not viewed as 'just being there'. Adults who have increased their own understanding of living things claim to have developed a completely different perspective on the natural world.

Focusing children's attention

Before any practical work is carried out within an environmental area, it is very valuable to provide children with a focus for their investigations. Without this, the environment presents a mass of bewildering information which can seem chaotic. We need to be clear about what we want the children to explore. Even when this is made very clear, children often divert to other investigations, as in the case of a very well prepared teacher, proud of the way in which her class were observing a grassy area for the various forms of life. A particular group, who had previously shown no interest, were totally engrossed in their task and with a congratulatory pat on the back, the teacher went forth to ask a few open-ended questions. On close inspection, it was obvious that the group had developed an alternative line of enquiry and were trying to set fire to the grass using the sun and a magnifying glass.

Close to home

Children at Key Stage 1 can learn about many aspects of living things by studying small areas close to home, e.g. by studying the variety of life in a school pond, under stones or on plants growing nearby. Every living thing needs its own particular niche and children of this age can readily identify with a need for a home of their own.

Using 'homes' as a starting point (where we live, what we need) can encourage young children to observe not only the form and function of living things but also how they are adapted to their own particular place in the wider environment. A wide range of adaptation can be noted by studying some of the more commonly found animals such as earthworms, snails, spiders, pond skaters, diving beetles and birds. This also allows the idea of characteristics of living things to be explored further. Table 2.14 shows a grid that could be used for this kind of study.

Table 2.14 Grid for exploring environmental concepts in living things

	Where found	*Shape and form*	*Movement*	*Adaptation*
Earthworm				
Snail				
Spider				
Pond skater				
Diving beetle				
Bird				

Modelling the environment

It is possible to explore the area of homes and shelter in the classroom before moving outside. Sand and water trays can be used to model homes, drawing both on children's knowledge and imagination. Sand trays can be used to explore wet and dry areas, shady areas and burrows, etc. Creating a three-dimensional space can be the beginnings of thinking about the environment in the same way. Providing appropriate material will enable children to make various models which incorporate the idea of homes in the ground, under stones, on leaves and in trees.

A water tray can be used as a model for a pond, allowing children to develop ideas of surface, middle and bottom areas. Through the use of simple reference books, children can become familiar not only with the types of organisms living in ponds but also where they are to be found within the pond. Teachers have effectively used a hoop to develop a pond mobile with children providing drawings of organisms to be found at different levels within the water, promoting a sense of a three-dimensional system containing defined areas. A plastic fish tank can be very useful for modelling a pond environment. (Work of this kind also allows children to develop their ideas of floating and sinking which links very closely with animal movement in water, e.g. use of air bubbles, surface skin, movement through water.)

The following case study illustrates how a teacher provided children in Year 1 with opportunities to develop their conceptual understanding together with their investigative skills.

'Minibeasts' is a popular area of investigation for children of primary age range. Looking at 'creatures' through a magnifier and finding out how they carry out certain life processes holds children's interest for prolonged periods. It is possible to differentiate the complexity of approach, encouraging simple investigative skills at Key Stage 1 and moving on to develop greater precision with increased complexity of experimental design and appropriate recording techniques in Key Stage 2.

The school was fortunate enough to have access to its own nature garden; a safe area for the children to explore. The following objectives were identified:

1. to encourage children to observe and identify 'minibeasts';
2. to develop children's investigative skills with particular reference to observation of similarities and differences, form and function;
3. to develop children's understanding of the concept 'habitat';
4. to develop children's skills of communication, together with vocabulary associated with the form and function of minibeasts.

Sequence of activities

Discussion
The class teacher introduced the area of minibeasts to the children by initiating a whole class discussion on 'What kinds of animals do you think we shall find in the school garden?' This gave the teacher the opportunity to assess the children's present knowledge and expectations. The children were able to share their ideas and a list of minibeasts was made for them.

Observation
The class teacher provided the children with large laminated photographs of the 'minibeasts' to be found in the school garden together with labels. The children were given time to observe the detail of the photographs which were laid out in the nature area as an aid for future identification.

Further discussion
Further discussion enabled the class teacher to explore the idea of animal homes and where they were likely to find them.

Collection
Using plastic petri dishes, paint brushes and plastic magnifiers, the children collected 'minibeasts' in the nature garden, paying particular attention to where they found them. The previous period of discussion enabled the children to be more confident in their search for a range of 'habitats'.

Observation/Identification
Paper and drawing materials were taken into the garden area so that the children could make detailed drawings of the minibeasts. The class teacher used open questioning to encourage greater depth of observation, for example: 'How many legs can you see? How do you think it moves?' Comparing the animals with the laminated photographs enabled them to name and identify them.

Recording
Returning to the classroom the children were asked 'Where did you find your minibeasts?' The class came up with a number of answers which were incorporated into a very simple grid entitled 'Where do minibeasts live?' (see Table 2.15).

Table 2.15 'Where do minibeasts live?' (grid for minibeasts' habitats)

	on the ground	*under-ground*	*under a rock*	*under piece of wood*	*on a leaf*	*on a stem*
worm						
ant						
snail						
spider						
woodlouse						
butterfly						

Sorting/Classifying
To develop their sorting skills and their ability to recognise similarities and differences, the children were asked to tell the teacher how the animal moved. They came up with a number of ideas:

- it wriggles
- it flies
- it runs
- it slides.

The children then sorted these animals into groups (Figure 2.25) using these categories.

With the group that runs, the children were encouraged to see how they could separate them. For example:

the spider has eight legs and spins a web;
the woodlouse has lots of legs;
the ant has six legs.

The children made a set of cards to go with the minibeasts and began to make an extremely simple key as shown in Figure 2.26. They had already had experience of sorting and grouping in other locations.

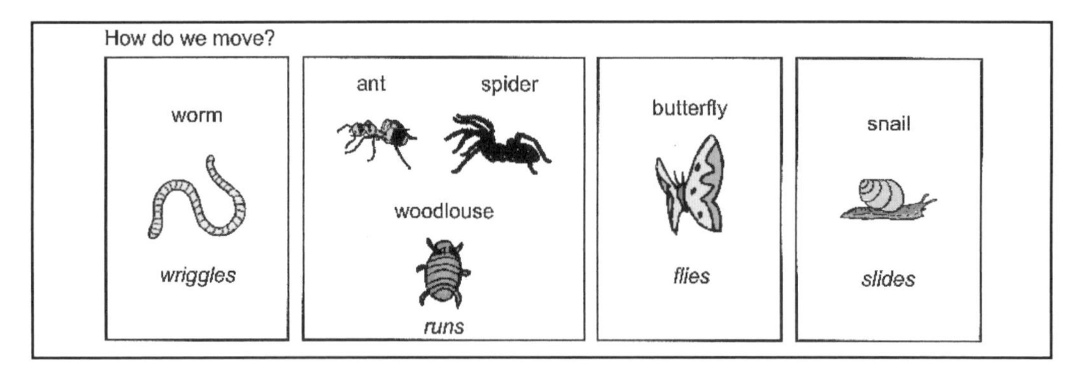

Figure 2.25 Sorting animals by their type of movement

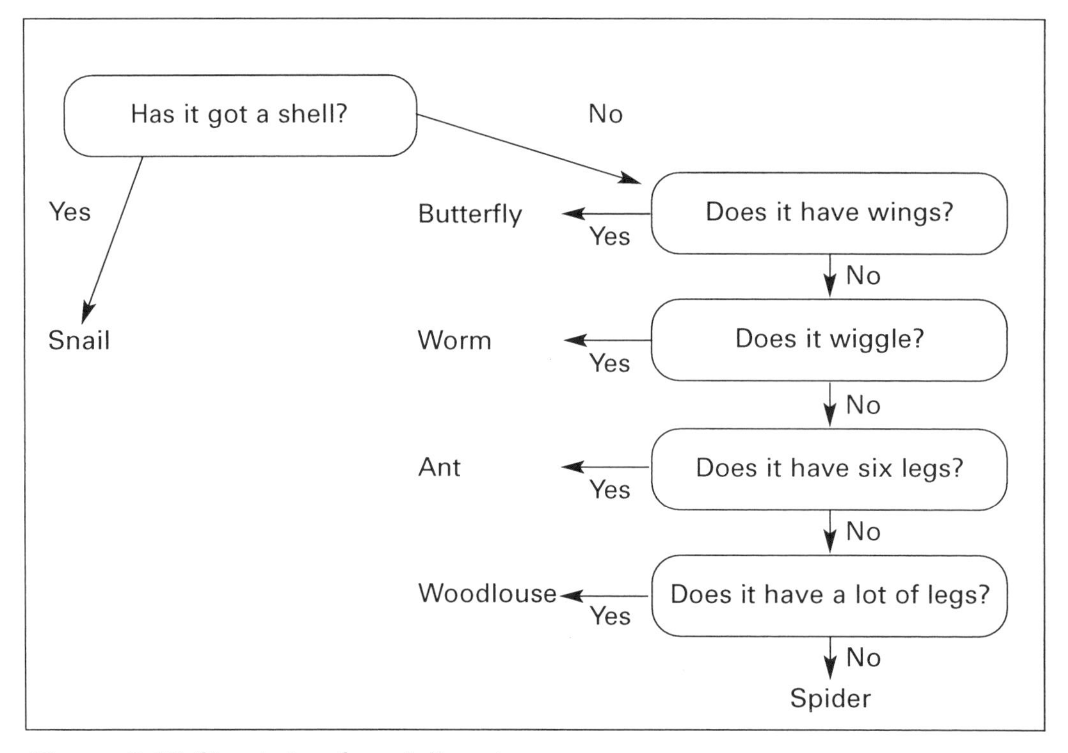

Figure 2.26 Simple key for minibeasts

Further investigation

The children and teacher set up a wormery in the classroom so that they could watch the worms burrowing and taking leaves from the surface down into the soil.

Recording

In groups, the children made large-scale models of their minibeasts, which encouraged further observation and discussion. The papier mâché minibeasts made a lovely display hanging from the ceiling.

Minibeast homes

Using their observation of minibeasts within the nature garden, children were given the task of using a large cardboard box to model a home for their minibeasts. The home was designed according to where the minibeast had been found (i.e. woodlouse under some wood, butterfly on a leaf). The children designed and made the collection of habitats using crêpe paper and tissue paper, junk modelling material, paint, etc. This provided reinforcement of the idea of many habitats within one environment.

The children helped to make a wall display of extra large minibeasts together with their homes, a useful way to reinforce ideas. They also produced home-made books, each providing information and drawings on one particular minibeast. Children included their questions about minibeasts, together with their ideas on how minibeasts move and what they like to eat.

Progression in children's understanding

The investigation carried out by Year 1 children could provide the basis for exploration with older children, building on earlier experiences and understanding. For example, the making of a wormery could lead to exploring the concept of food chains and interdependence.

At Key Stage 2, the environment can be investigated in greater detail; a more complex system of classification can be developed and environmental conditions can be taken into account. Conditions such as humidity and temperature can be explored together using more sophisticated sampling and collecting techniques.

As children develop their ideas of light, dark, shade, damp, dry and temperature differences, a much more detailed exploration can be carried out. Children can be encouraged to develop the skill of mapping an area and identifying a range of habitats within an ecosystem. Animal populations may be calculated using a variety of techniques and progressively more complex recording methods can be explored. An environmental study at Key Stage 2 should also allow children to explore the validity of their sampling and measuring techniques.

Encouraging the use of questioning can help children to develop the investigative skills of identifying variables and providing consistency in their approach, e.g.:

- What are we trying to find out?
- What do we need to measure?
- How shall we measure?
- How often?
- At what times?
- How many results do we need?
- How do we record them?
- What do our results tell us?

A simple example of an investigation which can encourage the above aspects is 'Where do we find snails?' This requires children to observe snails in different parts of the school grounds. It soon becomes apparent that snails are usually found in shady, damp conditions. A snail count will show that more snails are found moving about on wet, cool days rather than dry, warm days. A simple investigation of this could be carried out noting weather conditions and number of snails counted as shown in Table 2.16 and Figure 2.27. Although this is a fairly simple exploration, it does provide children with the opportunity to observe patterns in results and consider appropriate methods of recording.

Table 2.16 Snail movement and weather conditions

Day	*Number of snails moving about*	*Weather conditions*
Monday	10	Wet
Tuesday	9	Wet
Wednesday	3	Dry
Thursday	2	Dry
Friday	3	Dry

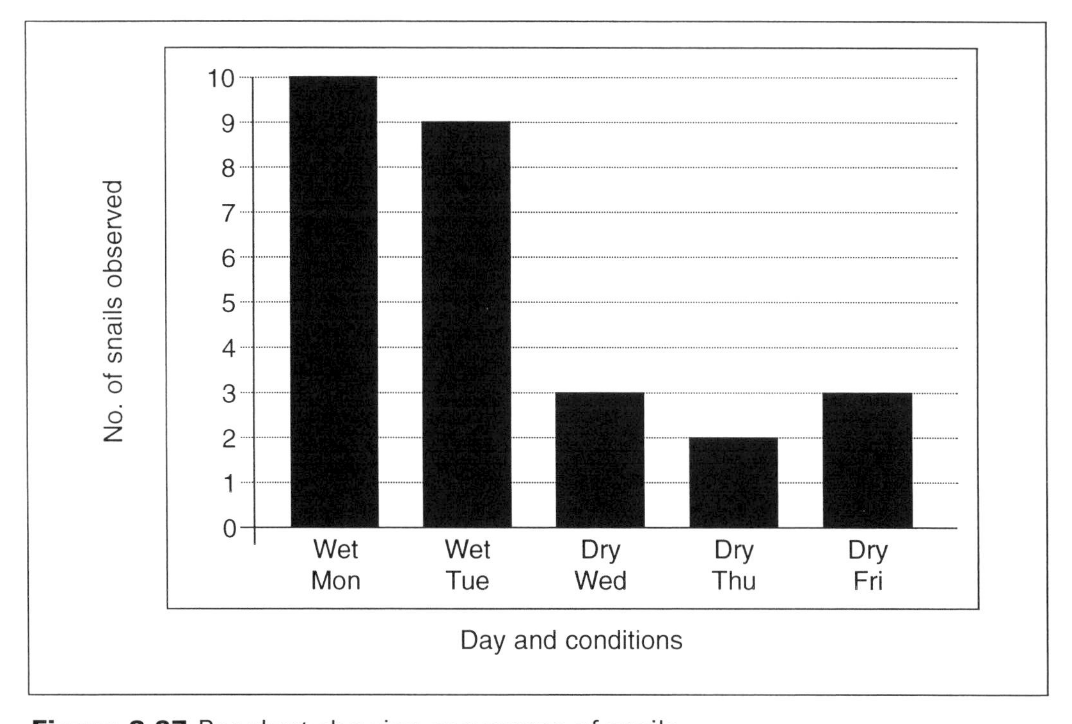

Figure 2.27 Bar chart showing occurrence of snails

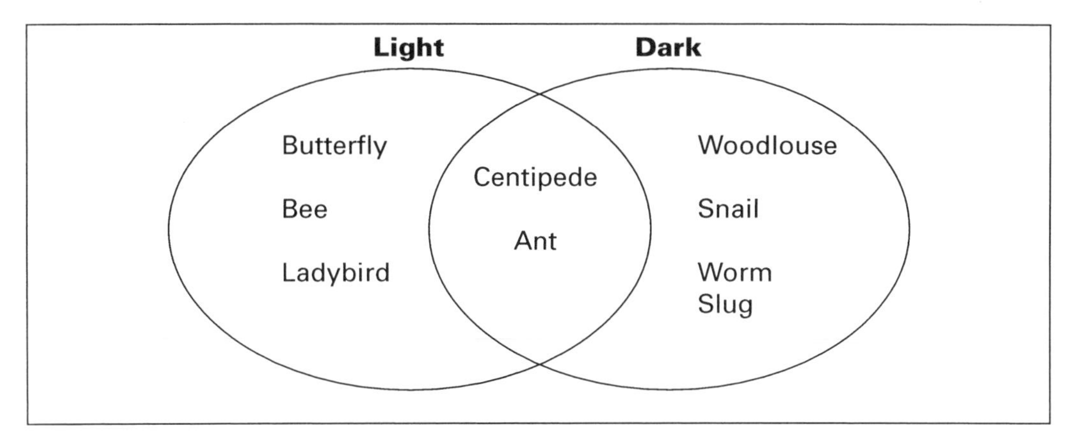

Figure 2.28 Two sets showing invertebrates found in the light and dark

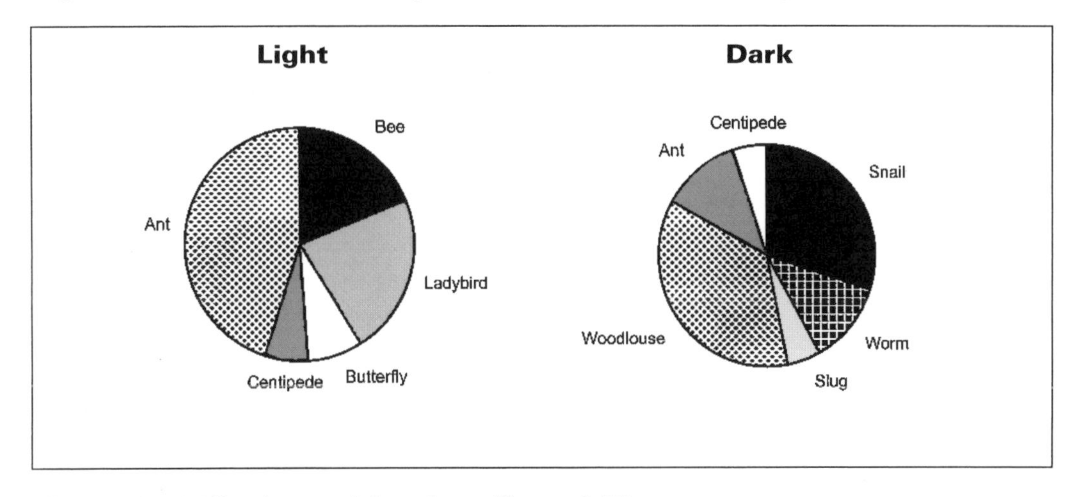

Figure 2.29 Pie charts of data from Figure 2.28

A more complex approach to sampling and investigative techniques can be developed when children are encouraged to note conditions in which several different invertebrates are found within the school grounds. An example is given in Figure 2.28. This type of investigation could also promote the use of more advanced methods of recording, perhaps creating pie charts to show the relative number (in this case, of each organism; see Figure 2.29). This investigation could also be used to explore:

- classification
- adaptation
- food chains
- reproduction life cycles
- movement (locomotion).

Studying a group of minibeasts provides children with the opportunity to explore the form and function of living things in relation to the environment. A range of adaptations for locomotion, feeding habits and life cycle can be considered.

At Key Stage 2, children can be allowed to investigate a more complex ecosystem such as a pond (see safety note at the end of chapter). Here, a range of habitats can be explored with children noting the incidence of particular species at different depths and in different zones (e.g. see Figure 2.30). This will provide many opportunities for increasing children's understanding of sampling techniques, observation and classification, and observing and measuring environmental conditions, as well as developing their concepts of interdependence, habitat, food chains and adaptation.

At Key Stage 2, progress in classroom modelling (e.g. a pond) can involve children in considering light and shade, damp and dry, and temperature differences. Before looking at environmental conditions which may affect the incidence of plants and animals to be found in and around a pond, it is valuable to explore environmental conditions from a number of perspectives. For example, children can observe a variety of areas within a playground and create a simple environmental map showing the range of conditions, before moving further afield.

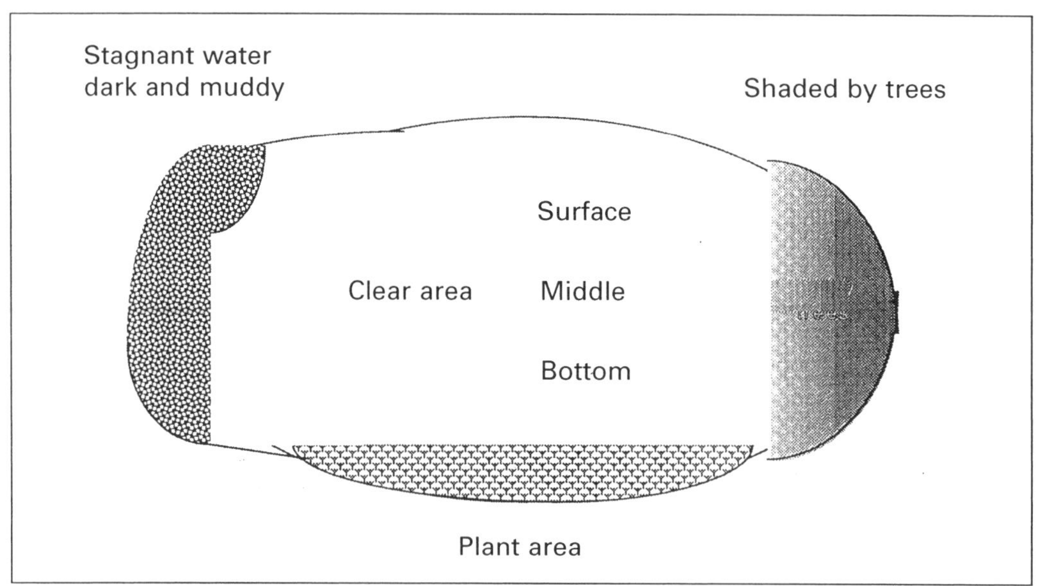

Figure 2.30 Study of a pond

THE INTERDEPENDENCE OF LIVING THINGS

Understanding the concept of interdependence

To understand the way in which ecosystems operate, it is important to consider the concept of interdependence. Underlying all investigation of the natural world is the interdependence of living things. All living things live within a balanced interdependent system with factors operating to maintain an equilibrium. This balance is the result of millions of years of a changing physical environment where survival of the fittest takes place. Animals or plants not adapted to survive when conditions change become extinct. This loss is still taking place today as environmental conditions are changed by the activities of the industrial and technological age.

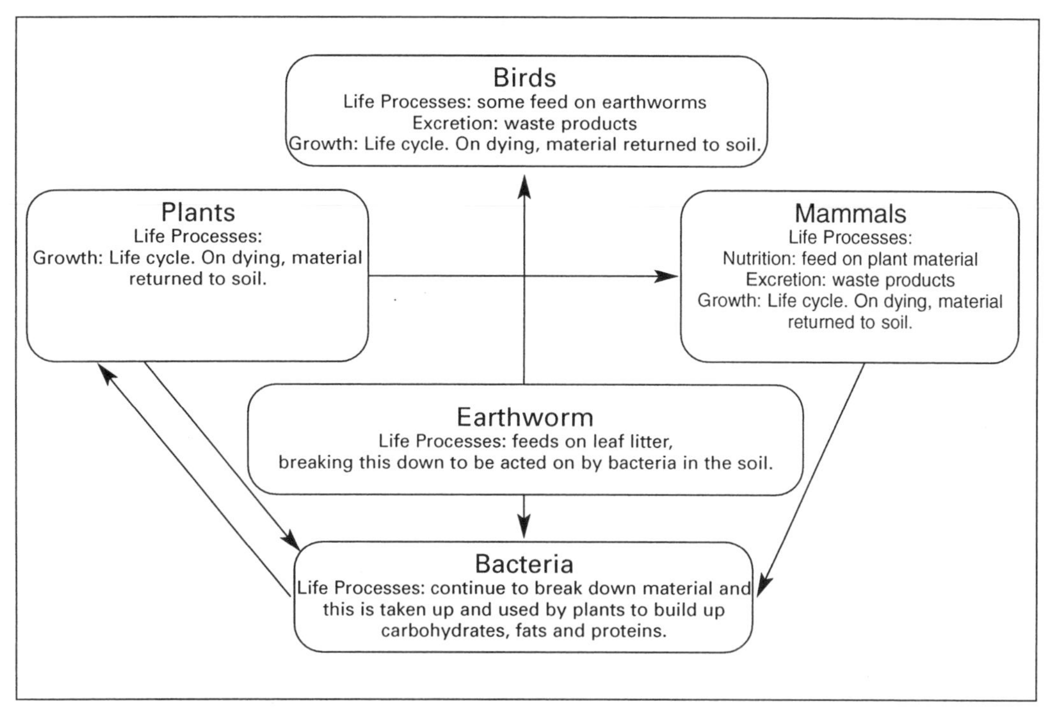

Figure 2.31 Interdependence between plants and animals

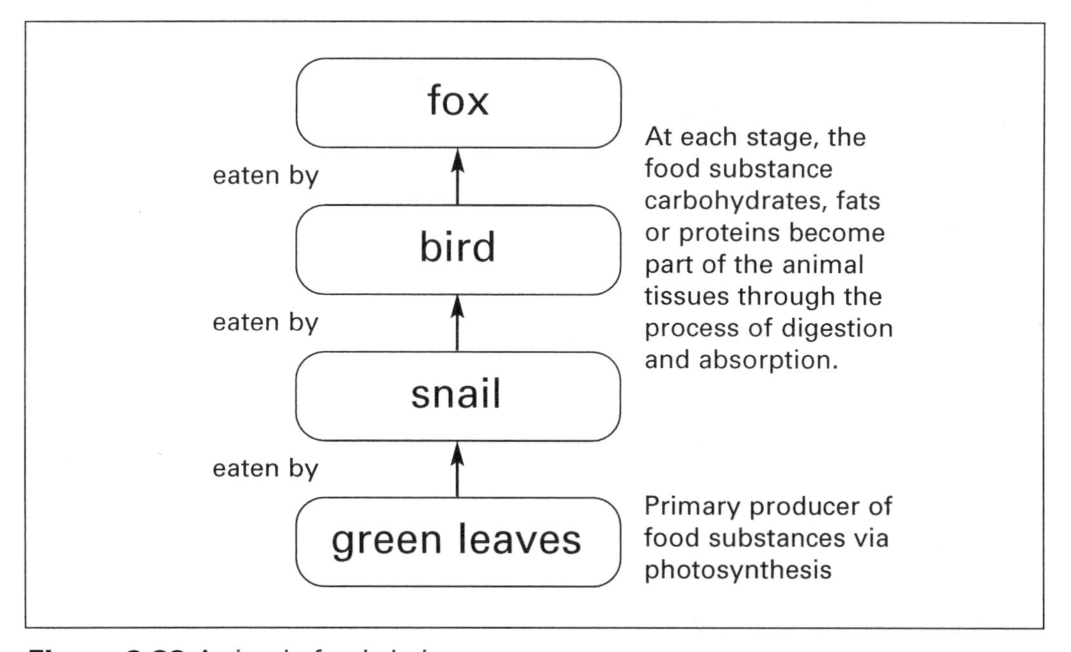

Figure 2.32 A simple food chain

We can take an extremely simple and commonplace example to illustrate the concept of interdependence; looking at the way in which the earthworm is part of a much more complex, interdependent system (Figure 2.31). The earthworm provides aeration of the soil, breakdown of organic material and is itself part of a more complex feeding cycle.

Living organisms are interdependent in numerous ways. Plants, as the primary producers, are able to build up food substances from simple materials, such as carbon dioxide and water. These substances are a source of nutrition for numerous other living organisms, which feed directly on plant material. Members of all major animal groups consume plant material, which through the process of digestion and absorption is built up into the animal's body. The interrelationship in terms of nutrition between plants and animals is called a food chain (Figure 2.32).

Animals which eat only plants are termed **herbivores**. Animals which eat only other animals are termed **carnivores**. Animals which eat a mixture of plants and animals are termed **omnivores**. All animals ultimately depend on green plants.

Pyramid of numbers

When looking at any food chain, it is usually very apparent that at the bottom of the chain the organisms are extremely numerous. As we move through each feeding level, the numbers reduce, giving a pyramid effect (Figure 2.33).

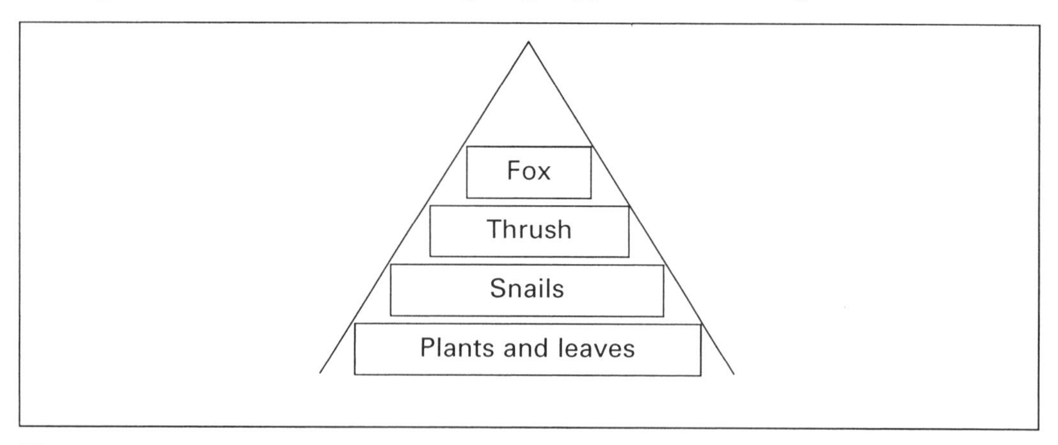

Figure 2.33 A pyramid of numbers

Homes and shelter

Another way in which living organisms develop an interdependent relationship is where one organism provides a home or shelter for another or, in the case of some plants, actual physical support. The detailed study of a well-established tree, such as an oak, shows clearly how many different areas of it are colonised by other living organisms (see Figure 2.34). All aspects of the tree (including leaves, stems, branches, bark, roots and leaf litter) provides numerous homes for a variety of species.

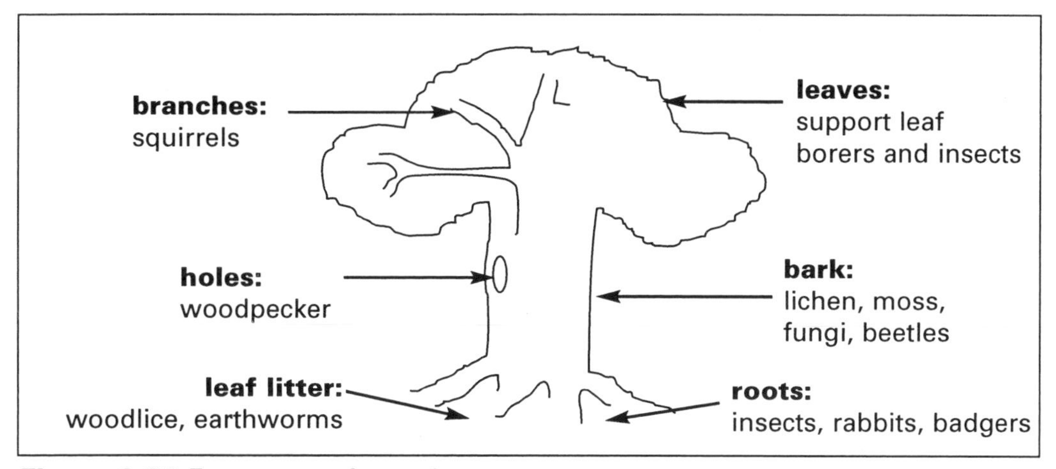

Figure 2.34 Ecosystem of an oak tree

Children's understanding of interdependence

Children at Key Stages 1 and 2 have difficulty in developing realistic ideas about interdependency as this requires not only an understanding of methods of obtaining food, but also an understanding of the general needs of a wide range of living organisms. The idea of decay and recycling also plays a part in the understanding of this concept, together with interrelationships to be found within an ecosystem. Children at Key Stages 1 and 2 often view living organisms as individuals without having a clear understanding of their relationship with other aspects of the ecosystem.

At Key Stage 2, children begin to develop an understanding of food chains and the role of plants as primary producers. A clear understanding of this, however, requires some knowledge of photosynthesis. At this stage, children are able to understand that organisms (with the exception of green plants) depend on other organisms for food and in some cases shelter.

Teaching about interdependence

At Key Stage 1, an exploration of 'What do we eat?' can develop children's understanding of where our food comes from. A simple survey of favourite foods can lead to children realising that much of our diet is dependent on plants, for example baked beans, cereal, tomato ketchup, fruit and vegetables.

By referring to simple reference books and by growing their own seeds, children can become aware of the origins of particular food materials. A variety of suitable plants can be grown in a fairly small space (refer to health and safety guidelines at the end of chapter for suitable plants).

The story of the origin of many food substances can be explored so that children become aware of how many plants and animals we are dependent upon. At Key Stage 2, children can gain information not only from resource books, but also from any local source of food production in order to gain an understanding of the sequence of events from growing, gathering, packaging and eating. Exploring materials also raises the idea of interdependency in terms of:

- Where do silk and wool come from?
- Which natural materials do we use in our home or at school?
- Where do they come from?

At Key Stage 2, exploration of a variety of habitats can reinforce the idea of interdependence in terms of providing food and shelter. Exploration of a single tree can provide many examples of homes for animals. As seen earlier in this chapter, the study of a pond clearly illustrates the interrelationship between plants, animals and the environment. At a simple level, noting feeding patterns between living organisms found within a pond develops the idea of food chains and pyramids of numbers.

The idea of interdependence can be reinforced by the use of games. Children can create their own games where drawings of animals need to be matched not only to each other but also to plant materials to form food webs. The making of models or mobiles can also help children to develop their understanding about the interdependence to be found within food webs.

HUMAN EFFECTS ON THE ENVIRONMENT

Understanding human effects on the environment

One of the main ways in which an ecosystem can be put out of balance is by the activities of humans. There are numerous specific examples of how the pollution in rivers, by pesticides, and from human and chemical effluent have had a very particular (and often negative) effect on the living organisms within ecosystems.

A particular case is that of the insecticide dieldril used to kill wireworms and other insect pests in the soil. This chemical was used to coat seeds before planting in order to prevent insect attack. The small amount used would not in itself have harmed other animals; however, birds feeding on the seeds consumed large quantities. As a result of this high concentration of poison, thousands of birds (such as rooks, pheasants and pigeons) were killed, together with foxes and birds of prey which fed on them.

Another well-documented insecticide, DDT, also led on to very unforeseen results and destroyed the balance within the ecosystem. The chemical was administered to water or soil in specific areas to eradicate insect pests, particularly the larvae of the mosquito responsible for transmitting malaria. DDT, however, was later found to accumulate in the fatty tissues of animals and so built up within the bodies of organisms at different feeding levels within a food chain. This happened until on reaching the highest level mammals and birds, the DDT concentration was so toxic that large numbers of these organisms were poisoned.

Although modern pesticides are broken down slowly to form harmless substances, it is impossible to find a pesticide selective enough to kill only the specific pest. Damage to other beneficial species (such as insect pollinators or necessary predators) usually occurs (see Figure 2.35).

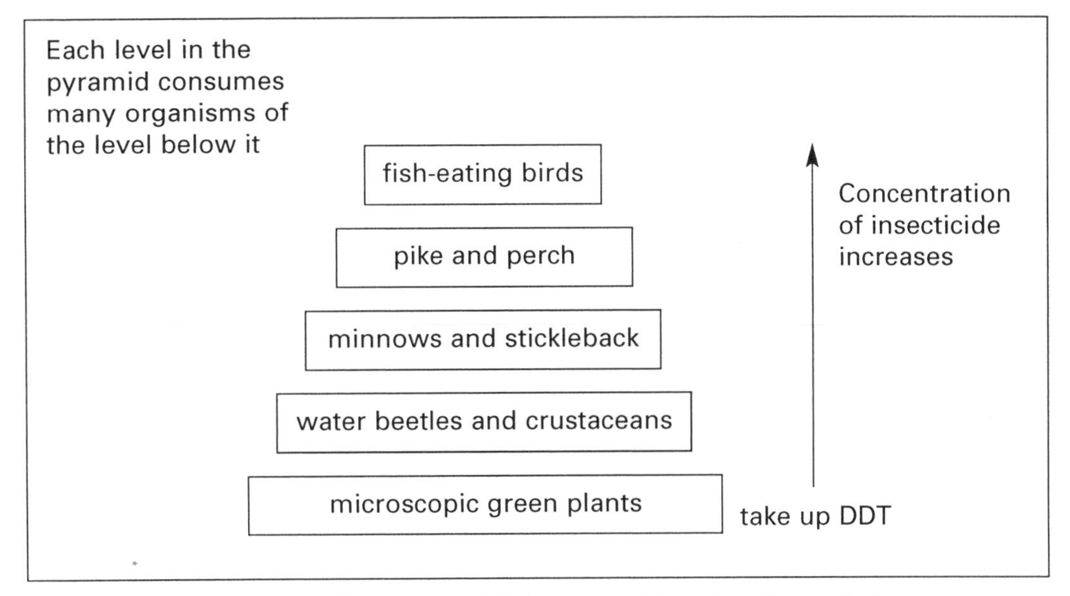

Figure 2.35 Example of how a pesticide can build up in a food chain

Many aspects of the modern world have implications for the balance of ecosystems and the living organisms within them. Effluents from chemical works, oil refineries and steel plants may be a source of pollution not only to rivers but also to land. Road developments which remove sections of woodland may also create an imbalance within the ecosystem. The destruction or reduction of one particular species can result in increased numbers of other organisms as the balance between predator and prey is destroyed. Every living thing within an ecosystem is held within a fine balance and any changes within the population may have far-reaching effects.

Children's understanding of human effects on the environment

Children have a somewhat limited understanding of the extent of human effects on the environment. They are often familiar with the damage caused to the rain forest and the idea of air pollution together with other information gained through the media. However, at Key Stages 1 and 2 children tend to have isolated pockets of information about human effects on the environment, without understanding their widespread results.

Teaching about human effects on the environment

Although much of a child's understanding of human effects on the environment will develop during the secondary years as they become increasingly capable of understanding the complex interrelationships of ecosystems, it is important to develop the early building blocks for later development.

Where do we start

At primary level, it is valuable for children to explore their immediate environment before looking at more complex issues further afield. The RSPCA education department produce excellent materials aimed at primary school children which can be used to explore some of the environmental issues associated with human activity. A variety of teaching materials are available including videos to promote children's empathy with animals and their needs and animal care work sheets which provide opportunities to explore the relationship between aspects of life processes and the needs of animals. These materials help children to develop their understanding further.

The RSPCA are involved in teacher education and a particularly interesting area which they explore is the way in which our household rubbish can kill or damage animals in the wild when taken to waste tips or dumped on the side of the road. Without us realising it, everyday household rubbish such as plastic bags or broken glass can maim and kill animals. Plastic can holders can become trapped around an animal's neck, balloons may be swallowed, and cans and pots can cause an animal to be trapped by its head. Using some of these very direct ideas with children can easily lead on to ways in which they can prevent this happening.

At Key Stage 2, children can employ role play to develop their understanding of human effects on the environment. Presenting children with a specific case study, such as a plan to develop a by-pass through a woodland area, provides children with the opportunity to present the case for and against the proposal. Providing arguments for and against involves children not only in role play but also in researching, presenting and communicating, taking into account a number of viewpoints.

Essentially any study of human effects on the environment brings together many areas of understanding including:

- life processes and animal needs;
- interdependency;
- adaptation;
- ecosystems and the need for balance.

SOME HEALTH AND SAFETY ISSUES

The ASE booklet *Be Safe* (see Appendix) together with school LEA Health and Safety Guidelines should always be referred to before undertaking investigations with living things and the environment. Plants and animals brought into the classroom must conform to LEA Health and Safety Guidelines. Be aware of protected species and poisonous plant materials such as red kidney beans and castor oil seeds. Children should be made aware of the danger posed by some plants in the environment.

Disposable gloves should be worn for pond dipping as water can carry Weil's disease. They should also be worn when handling soil. Any environment should be checked for safety hazards before being investigated.

CHAPTER 3

Materials and Their Properties

INTRODUCTION

The opportunities for both the pupil and the teacher for the exploration of materials in the primary school are all-encompassing. For whatever the child is investigating, whether it be a specifically scientific topic like forces or electricity, or some other curricula topic where any tangibles are concerned, then 'materials' will be involved, simply because the Earth and everything on it are made of materials. 'Materials' includes all of the materials of which things are composed, i.e. natural materials, such as rocks and air; materials that have been made from other materials, such as glass; and those which make up living things, such as teeth and bones.

The term 'materials' is often used in a rather limited sense by children for things like fabrics and building materials. Teachers need to develop and extend the children's perception of the word 'material' through the exploration of a large range of experiences with materials which will extend their understanding and allow them to differentiate between an object and the material of which it is made. In scientific terms, the word 'material' is used widely to include liquids and gases as well as solid materials. There is a huge variety including naturally occurring materials like rock, wood and bone, traditional ones such as stone and glass, as well as modern materials including plastics. The creation and use of collections of materials in the classroom provides opportunities to investigate similarities and differences between materials. 'Feely bags' are particularly useful for encouraging children to consider the relative properties of materials (e.g. smoothness, hardness, etc.) without the distraction of actually seeing the object.

An example of how a teacher creates opportunities to develop and extend pupils' experiences is shown in this account of work on metals undertaken by a Year 4 class.

The class teacher had displayed a collection of metal objects which included paper clips and drawing pins, keys, kitchen utensils, tools, metal toys and coins. The children were encouraged to handle the objects and make observations. They commented on the texture of the metals, their comparative hardness, smoothness, colour, shininess, coldness and heaviness. In small groups of four, the children were asked to investigate just one of the properties of the metals that was of particular interest to them. One of the groups was interested in whether or not all of the metals would conduct electricity (a topic that had been visited earlier). A simple circuit incorporating a light bulb was constructed and the different metal objects were put into the circuit one at a time to see if the bulb would light. Thus the metals were grouped and common properties established.

Another group investigated whether the metals were attracted to magnets or not. Again properties common to some, but not all, were established. One of the groups chose to investigate the flexibility of the metals, having found out that the paper clips would bend much more easily than the coins. This involved the teacher and children in much discussion of how the relative thickness of each object would allow them to test fairly.

The children added to the collection with metal objects brought from home. Copper tubing, aluminium foil, lead, silver, brass, bronze, the liquid metal in a mercury thermometer and even some gold! This lead to discussion about why some of the metals had rusted and others had not. The combination of metal with other materials was considered, e.g. why was the copper wire in the circuit covered with plastic? This lead to the consideration of conductors and insulators, thus revisiting and extending the work on electricity. Links with history and geography as well as geology were made. Where are these metals found? How are they mined? How are they extracted and mixed together? Why are some metals chosen for particular purposes?

The children were having extremely valuable experiences of a variety of elements concerning metals – the mechanical properties of metals, the changes brought about by heating metals, making and obtaining metals and experience of metals and their uses – and all derived from the initial collection of metals. (Another case study relating to the nursery classroom can be found in the section 'Understanding heating and cooling'.)

THE STRUCTURE OF MATERIALS

Understanding solids, liquids and gases

It is very useful for both teachers and pupils to have some knowledge of the basic structure of solids, liquids and gases, in order to understand the behaviour of different materials.

All materials are made from minute particles called atoms, and these can join together in small or large groups to form molecules. Under the normal range of temperatures these molecules are constantly vibrating rapidly. It is the way in which they are arranged and how close together they are that will determine whether a material is a solid, a liquid or a gas.

- There are only minute gaps between the molecules in solids. The molecules are held rigidly together (or bonded) and they vibrate about a fixed point.
- The gaps between the molecules in liquids are further apart and this allows the molecules to slide over one another so that the material can flow.
- The gaps between the molecules in gases are even further apart allowing them to move freely.

This form of movement is a manifestation of the kinetic energy – the energy that moving things have – of the atoms. The amount of energy will determine whether the substance exists in a solid, a liquid or a gas state. If the energy is increased sufficiently in a solid (for example when chocolate is heated), then the bonds between the molecules break down and the particles can move more freely in all directions. The solid becomes a liquid. If the amount of energy is decreased, then the liquid will become solid.

Particular points to note are:

- atoms are the building blocks of life;
- many substances can exist in a solid, liquid or gas form;
- something is still the same substance, even if it has changed from one state to another.

Children's understanding of solids, liquids and gases

Research carried out by the Primary SPACE team (Science Process and Concept Exploration 1991) indicates that children at both Key Stages 1 and 2 have a much clearer understanding of the solidness of a substance and the liquidness of a substance than they do of the nature of a gaseous substance.

Most young children do not think of gases as being materials at all. They associate the existence of air only with moving air, although they understand that air enables us to live. However, they believe that 'gas' is dangerous. It is therefore not surprising that when asked 'What is in the other half of a glass that is half-filled with orange juice?', almost all children will answer 'Nothing'. For almost all of the gases that children will have had experience of, even if they have been unaware of it, will have been invisible. Most of those gases which are coloured are also poisonous.

Very young children will usually view solids as being hard, strong and rigid. The use of expressions such as 'solid as a rock' help to reinforce this erroneous view. It is important to provide children with as wide a variety of experience as possible, e.g. 'pour the flour and sugar into the bowl', in order to enable them to develop an understanding that solid materials can also be soft and non-rigid.

Liquids are usually discussed by children in terms of being watery and runny and materials that can be poured. There is a need to allow children to experience a variety of liquids with varying viscosity, comparing perhaps the behaviour of water and golden syrup. They also need the experience of pouring liquids to and from a variety of containers, in order that they can develop an understanding of the conservation of volume of liquids.

Teaching solids, liquids and gases

The class teacher is faced with the problem that easily constructed models showing the behaviour of molecules for solids and liquids, will be much more convincing than for gases. Using identical containers to hold materials it can be seen that solids such as blocks of wood and rocks will retain their own shape and will not spread out to fill up any spaces. The molecules of which they are made are bonded together and not moving apart from one another. Liquids, whether they are 'thick' such as syrup or 'thin' such as water, will take up the shape of the container, filling in the gaps. The molecules in these substances have larger gaps between them and so they can move around to fill in the spaces. Gases will fill whatever shape and size of container they are in, because the spaces between the molecules are large, enabling the molecules to move around more freely.

Three containers holding different quantities of marbles can be used to illustrate these principles (Figure 3.1). One container filled to capacity represents a solid, a second half-filled container represents a liquid, and a third, with only a few marbles in it, represents a gas. When the containers are shaken up the marbles in the filled jar cannot move around very much: they 'stay solid'. In the half-filled jar the marbles move around and over one another, representing the flowing movement of a liquid. In the third jar, the marbles with plenty of space around them move around even more freely, behaving like the molecules in a gas.

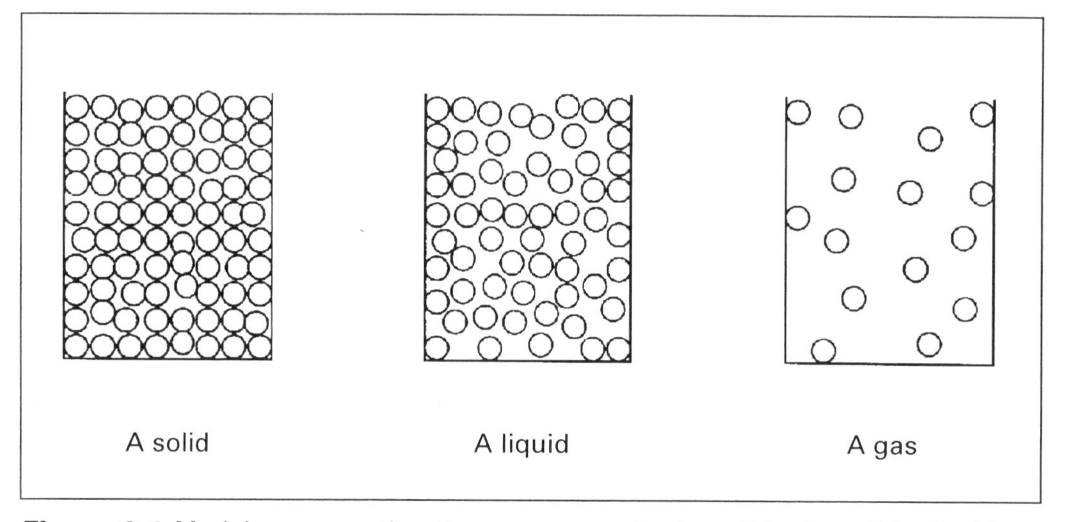

Figure 3.1 Models representing the arrangements of particles in solids, liquids and gases, using polystyrene balls in lidded jars

Young children can be encouraged to experience the substance of air around them, by having to push their way through it, holding a large sheet of cardboard in front of them. This effort could be compared with the relative ease of pushing without the extra surface area. The cardboard packaging from large domestic appliances could be used, as shown in Figure 3.2. (See also Figure 3.5 for increasing awareness of the 'substance of air'.) The way gases fill the whole of a container, in this case the classroom space, might be shown by taking the lid off a bottle of perfume and detecting its spread. How far does the perfume spread around the room?

Some of the properties of a gas might be shown using a model Cartesian diver (see Figure 3.3) or a plastic syringe. It is relatively easy to push the plunger into the syringe filled with air when a finger has blocked the escape route, squashing the air into a smaller space. When the plunger is released, it can be seen to return to its original position as the air 'pushes back'. If the syringe is filled with water and the tip is blocked up, then it is impossible to push the plunger in. The water cannot be compressed.

The Cartesian diver behaves in the same way. When the bottle is unsqueezed the diver, represented by the dropper, remains at the top of the bottle. When the bottle is squeezed, the only material that can be squashed up within the bottle is the air in the dropper. The water is forced to push on the air in the dropper, compressing it into a smaller space. Water enters the dropper to take up the extra space, the dropper becomes heavier and sinks to the bottom of the bottle.

Figure 3.2 A child running against the air holding a shield of cardboard

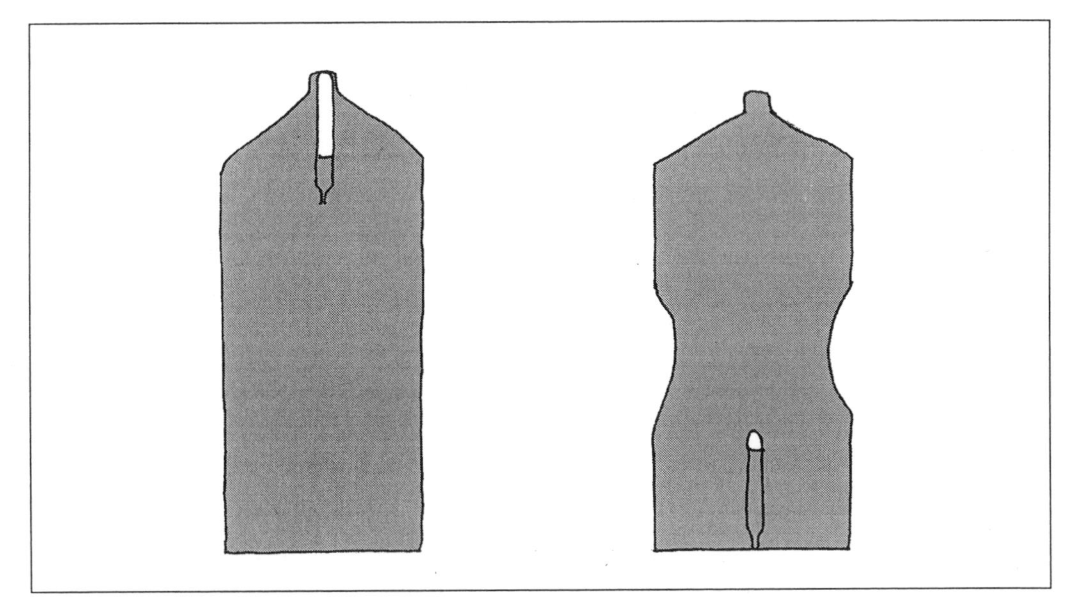

Figure 3.3 Showing the compressibility of air using a Cartesian diver

Children can also explore 'pouring' air bubbles under water from one container to another (Figure 3.4).

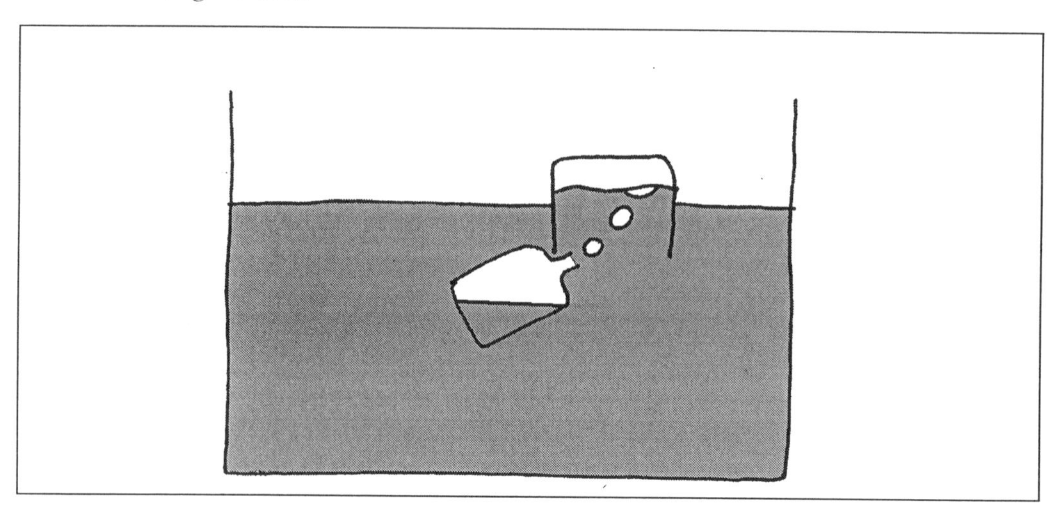

Figure 3.4 The air in a jar can be poured into the beaker of water, filling it with air

Weighing the air in balloons by balancing an air-filled balloon against an empty balloon may give children some evidence that air has 'substance' (Figure 3.5). Balloons of different shapes and sizes can be filled with air and investigated. The effect of helium-filled balloons could be observed, thus introducing the idea that gases can be comparatively heavy or light.

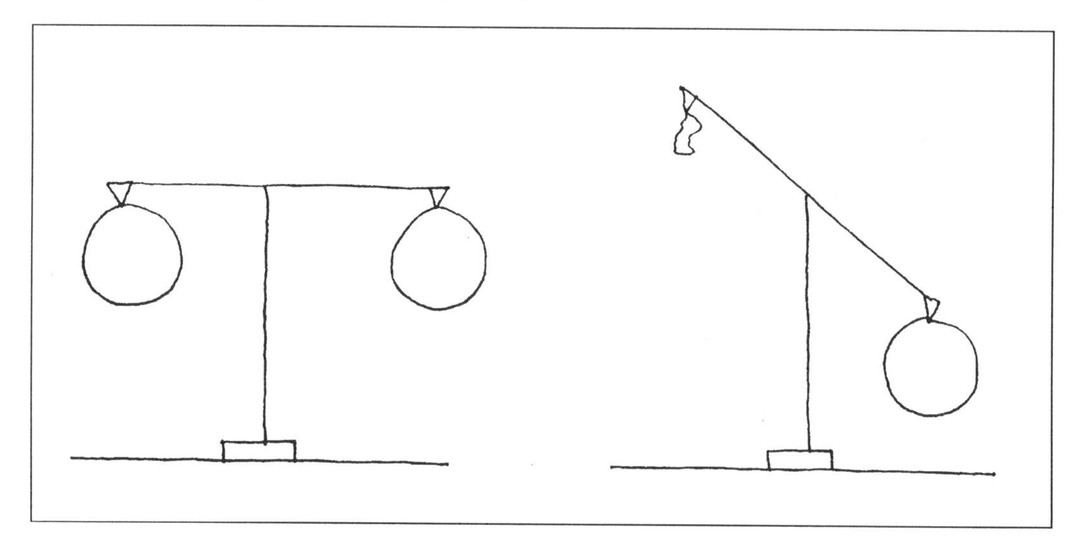

Figure 3.5 Does air have weight? The inflated balloons are balanced. When one balloon is burst, the balloon with air in it is heavier and the balance alters

Children need experience not only of the three states but also of the enormous variation within each. They need to move from classifying a wide range of materials as either solid, liquid or gas to classifying materials which cannot be so easily defined; materials such as jelly, Silly Putty and clay and fine powders such as talc would be useful examples. Children can be encouraged to build up an agreed set of criteria for the three states. For example, some solids such as Plasticine and putty can be squashed and stretched; others like rocks can be tested for their hardness. Liquids vary in viscosity, and children can experience the relative difficulties involved in pouring liquids like syrup, oil, shampoo, fruit juice and water.

PHYSICAL CHANGES

Understanding heating and cooling

The heating of materials in primary school is usually linked to cookery. Although many children will have taken part in the preparations for baking bread and cakes, etc., experience will usually have been limited to weighing and mixing the ingredients. The physical and chemical changes brought about through heating are often hidden from view because they are taking place inside the cooker.

By heating and cooling everyday materials in open aluminium dishes, as shown in Figure 3.6, children will be able to observe at first hand many physical changes. Some of the changes observed will be changes in state, e.g. chocolate melting when heated and becoming liquid, and then becoming a solid again as it cools. This change is reversible. Some materials change irreversibly when they are heated; e.g. sugar crystals changing from a white solid to a yellow and then golden brown liquid and finally a blackened solid is an irreversible change. Children will also be able to heat small quantities of water using this apparatus and use this to investigate dissolving materials like salt and sugar.

When a material is heated, the molecules from which it is made move faster. The effect of this is to increase the gaps between the molecules and consequently the material expands. If the material is a gas, e.g. air in a balloon, then as the air is heated it will rise up as the gaps between the molecules increase and the molecules move more violently. They take up more space, therefore increasing their volume. When the material is cooled the movement is less violent and the gaps close and the material contracts. In the case of the balloon, the air in the balloon will contract and the balloon will descend.

If the material being heated is a solid, then the closely bonded molecules will break down, enabling them to move freely over and between one another. The effect of this can be to cause the solid to melt and form a liquid, as do chocolate and jelly. These changes are reversible as long as the materials are not burnt. If the material is burnt then the changes are irreversible. This would be the case with wood, wax and natural gas. (See 'Burning' later in this chapter.)

If the material being heated is a liquid (e.g. water) then it will become a gas, evaporate, and as it cools it will condense to form a liquid again.

- A liquid material is evaporating slowly all the time to become a gas. A wet painting and a puddle will dry out as the water evaporates.

- When enough energy is transferred to a liquid substance, the process will speed up and the liquid will evaporate quickly – it will boil.

Children's understanding of heating materials

Children will accept that physical changes occur to many of the materials around them. They will be familiar with ice cream and chocolate melting, often at their inconvenience! The opportunity to observe other materials being heated under more controlled conditions will fascinate them. Young children do not usually distinguish between melting and dissolving. They often view melting as a gradual process that is similar to dissolving. Up until the age of 8 years, they will tend to focus only on the solute (see Glossary). For example, when sugar is added to water, children will say that it is the sugar that has changed by disappearing. They do not recognise that the water has also changed. Older children will suggest that the sugar has 'gone into tiny little bits'. Children also find it much easier to think of solids dissolving in water, but the idea that liquids and gases can also dissolve is less acceptable. Young children need encouragement and experiences to see that melting involves one substance and heating or warming up, and that dissolving involves two substances and does not necessarily involve heating.

Evaporation is a difficult idea for most children, but is especially so for the very young. A child aged about 5 or 6 will believe that when a wet surface dries, e.g. on a puddle or their newly finished painting, the water has disappeared completely. The same age group will know and accept that water boils away, although they will find it difficult to accept that the bubbles are gas escaping from the liquid. Some will comment upon steam misting up windows and know that they can draw on it and that their fingers will become wet. However, they will not always link this 'wetness' to the 'wetness' of the water that has been heated. When many children reach 9 and 10 years of age, they will suggest that the water has gone somewhere, although they are not sure where.

There is often confusion between the terms condensation, evaporation and boiling. Children need repeated experiences of these phenomena to understand that condensation is the reverse of evaporation and not the reverse of boiling.

Teaching about heating materials

There are obvious health and safety issues to be considered here. Figure 3.6 shows a suggested, safe method of heating. The organisation of the classroom is of great importance when working in this way. The teacher would certainly remain with the heating group at all times, with perhaps four children in the group for their first experience of heating and possibly up to six if they were more experienced. This is not only to ensure safety, but also to be on hand to draw and focus the children's attention and to ask appropriate questions. Issues relating to classroom organisation for primary science are discussed in Chapter 5.

The children should have any long hair tied back, and there should be the minimal amount of paper and other inflammables on the table. Probably the greatest danger comes from the children burning their fingers by touching the contents of the foil 'dishes' rather than the candle flame. To minimise burn injuries the children could hold a lolly stick at all times to protect the fingers if they did reach out, in addition to being reminded that they must not reach out to touch the materials.

The method of heating shown in Figure 3.6 can also be used to heat small quantities of water to investigate dissolving materials, making solutions and evaporation. Materials can also be burnt directly in the candle flame using metal tongs to hold the samples.

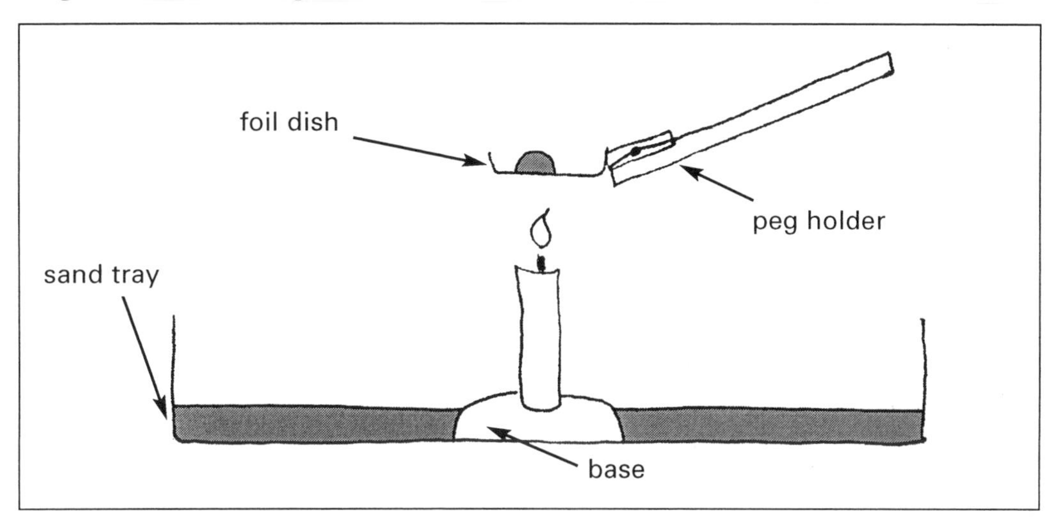

Figure 3.6 Heating materials in the classroom

Children should have the experience of heating a large variety of materials and repeating their observations where appropriate. The advantage of observing sugar melting and flour and water 'rising' first hand is invaluable, with great opportunities to develop observational skills.

Because children think that there is a loss of mass and weight when a substance melts there is plenty of scope for accurate measurements to be taken of materials before and after melting, if possible using electronic scales. Similarly, measurements taken of solids and the solvent before and after dissolving substances is useful in encouraging the idea that the substance has not disappeared. Regaining the sugar from the water through evaporation would also be helpful in reinforcing this concept.

Children could also be encouraged to use the most appropriate language when describing their observations. For example the explicit use of the words melt, dissolve and evaporate.

The children can be encouraged to use all of their senses when observing the effects of heating, including listening for gas production through the hissing and bubbling of materials, as well as smelling the gases. There are, however, significant problems with children tasting materials, and great care should be taken.

Health and safety note
Children must be encouraged to smell the materials by wafting the air above the materials towards their nose and not to inhale while their nose is over the material itself.

Many materials can be investigated, including cooking and eating chocolate, flour, bread, water, jelly, ice and wax. It is useful to experience dramatic changes as well as the less dramatic. Icing sugar and plain flour look very similar. However, when small quantities are heated the icing sugar will change dramatically, whereas the flour will change very little.

One nursery class teacher involved her children in heating by cooking a whole fish! A newly dug pond in the school grounds was the stimulus for them to investigate fish. The children were unsatisfied with the glimpses of the goldfish that they saw in the pond and wanted a much closer look. Catching an individual fish and putting it in a bowl gave a better view, but, of course, the fish kept moving and also could not be touched. The children were very curious to know what the scales and the fins felt like. Did the fish have a tongue? How squashy was its body? Did it have bones? When questioned, it became obvious that many of the children had not experienced eating whole fish, but only fish fingers.

The teacher brought a whole trout into the classroom and the children were allowed to examine it through touch, sight and smell. The scales, eyes, gills, mouth, relative firmness of the flesh, the weight, the flexibility of the body, etc., were observed with great enthusiasm and interest, the children were fascinated and the activity promoted many questions. Are their eyes like ours? Can they blink? Do they sleep? Does their colour change as they grow older? Can they hear? Where are their ears? Why don't they float? What is under the scales?

There were plenty of ideas to follow up. The teacher in this case, took up the last question. A section of skin was cut away to reveal the flesh which was observed. The almost transparent nature of the flesh surprised many of the children. Realising that it would be much easier to examine once the fish was cooked, the trout was pan fried whilst the children looked on. The irreversible changes in the appearance and texture of the fish were noticed and the fish was tasted by those who wanted to and the skeleton was revealed. Altogether a very rich, scientific experience.

MIXING AND SEPARATING

Understanding mixing and separating

Young children need little encouragement to mix things together. Considerable time is spent in the early years 'sorting out' the boxes of building and other materials which have been put together either intentionally or unintentionally. Sand in the sand tray can have materials in it sieved out and floating materials can be scooped from the surface of the water tank. Many materials are mixed together in bowls in the home corner for 'pretend' meals, as well as 'real' ingredients for 'real' cooking.

Through firsthand experience, children can mix together and then separate everyday materials, developing their understanding about many areas related to materials. A mixture is made when two or more substances are physically mixed together to make what appears to be a single substance while the constituents remain chemically discrete.

There are many combinations which can be investigated and it is useful to consider these in six broad areas:

- a solid mixed with another solid, e.g. salt mixed with sand, or pebbles mixed with wood chips;
- a solid mixed with a liquid, e.g. coffee granules or sugar mixed with hot water;
- a solid mixed with a gas, e.g. the smoke from a wood fire with the fine particles of wood in the air; also tobacco smoke and pollution produced by motor engines;
- a liquid mixed with another liquid, e.g. the droplets of fat in milk (this is known as an emulsion);
- a liquid mixed with a gas, e.g. hair spray and deodorants in aerosols, clouds and fog, and carbon dioxide in fizzy drinks;
- a liquid mixed with a porous solid, e.g. water in a bath sponge or a face flannel.

The manner of separating the materials that have been mixed will depend upon the properties of each of the materials in the mixture. For example, if one of the materials in the mixture is magnetic, perhaps if steel paper clips have been mixed with salt, then it would be possible to separate the two materials using a magnet. Equally, if one of the materials dissolves then the separation can take place through a solution being formed in warm water, and then the solid being evaporated out.

Solid-with-solid mixtures may be sieved if the particles are of sufficiently different sizes (see the section on rocks and soils). If the particles are very fine and do not dissolve, then they may be separated through sedimentation when the mixture has been shaken up with water and left to settle.

It may be useful to distinguish between a solution and a suspension. When a solid is mixed with a liquid, the mixture which is produced is either a solution or a suspension. Which, will depend on whether or not the solid will dissolve in the liquid. For example, if coffee granules are mixed with warm water then the mixture will be clear. If the coffee–water mixture is poured through a filter paper, no solid material will be filtered out. A solution has been produced. However, if a sample of soil is mixed with warm water the mixture is cloudy, and if passed through a filter paper a solid residue is left on the paper. This is a suspension.

Materials which form suspensions do so because they are small, insoluble solids. Materials which form solutions do so because they are made from chemical particles which are many thousands of times smaller than the smallest solid particle. When a solid dissolves it does so only at its surface. As a result, a single lump of solid will dissolve more slowly than the same amount in powdered form.

A solution is formed when a substance (a solute) dissolves in a liquid (a solvent). Water is the most frequently used solvent in the primary classroom. In order to separate the materials, the solute may be recovered by allowing the water to evaporate or speeding up the process by boiling. This can be achieved using the resources shown in Figure 3.6.

Gases, including air, can dissolve in water and other solvents. The temperature of the solvent effects the amount of gas that can dissolve. Whereas an increase in temperature of water will enable more solid material like salt and sugar to dissolve, the reverse is true for air. The cooler the water the more air is dissolved in it. When heating water the bubbles of air can be observed.

Children's ideas about mixing and separating

Children often refer to 'pure water' and 'pure air' without necessarily having an understanding of a single substance. They tend to think of 'pure' as meaning 'without having anything harmful in it', often thinking in terms of pollutants. They think of materials as being a single substance unless they can physically see several component parts, e.g. in a jar containing pebbles and wood shavings, or in a slice of fruit cake. It is useful to develop their understanding from the idea that objects are 'made of' materials. This can be extended and developed to pursue the 'What is it made of?' question and then to ask 'What are materials made of?'

Most children when asked the question, 'What happens when a solid, like sugar, is added to warm water?' will respond, 'it disappears'. Their observation has focused only on the change to the solid, rather than on any change to the liquid (see also section 'Physical changes').

Teaching mixing and separating

The resourcing for this area of materials is relatively simple. A selection of materials from the kitchen with a few additions from the garden should suffice. A consideration of the simple properties of the materials, such as whether they dissolve in water, float or are magnetic will widen the range of mixing and separating. Here are some examples:

- Nursery children are usually skilled at sieving out toys from sand. This can progress to sieving fine solids like salt or sand from pebbles, flour from marbles. Attention can be drawn to the size of grill in the sieve.
- Adding food colouring to the mixtures will encourage and stimulate younger children particularly.
- Small quantities of salt which have been mixed with warm water and dissolved can be left out on a saucer. When the water has evaporated the salt alone will be left.
- Lolly sticks, plastic spoons, ping pong balls, pebbles, shells, and any other materials that will not spoil by being immersed, can be put into bowls of water and then those materials that remain on the surface can be scooped out. (Any evidence of rusting, however, may be used to support work on chemical changes to materials, see next section.)
- Including iron and steel in a mix will allow for separation through the use of a magnet.

- Multiple mixes can be produced for older children, perhaps of wood chips, salt and stones and the question asked how can they be separated.
- Older children investigating mixtures in solution, can consider saturated solutions. How much salt can be dissolved in a limited amount of water? Can we go on adding more and more, or is there a limit? Consideration can be given to the temperature of the water. If the water is hotter can it dissolve more of the solid?
- Materials which remain obviously discrete such as oil and water are also useful for investigation.

Many of these investigations will link in with sinking and floating work as well as whether materials will or will not dissolve. Other useful materials to use may be cornflour, baking powder and sherbet. The last two will enable the children to investigate the production of a gas as the materials effervesce.

As discussed previously, the development of concepts relating to the properties of gases are the hardest to secure. Many young children think that air is made of one substance rather than being a mixture of gases. They do not realise the difference between air breathed in and air breathed out, for example.

Air mixed with water can be seen when water is boiled. Most children are familiar with the rush of gas escaping from a bottle or can of fizzy drink as the pressure is released. Their acceptance of these phenomena needs to be challenged by the teacher through appropriate questioning. What is inside these bubbles? Is it the liquid or is it something else? Links can be made with observations made of bubbles of gas being produced when heating materials such as sugar.

CHEMICAL CHANGES

Understanding chemical changes

Careful observation of some everyday materials in different situations will give good evidence of the chemical changes which materials undergo. Many of these changes are associated with cooking. Boiling an egg changes the proteins within it, causing it to coagulate and thus become bound together. Cake mixture also changes when heated: the proteins in the flour combine with water and become very elastic, giving the cake its spongy texture; baking powder produces bubbles of carbon dioxide which stretch out the elastic gluten. Sometimes when cooking, things do not always go strictly to plan, and we often say that the food is burnt!

Burning

This is the name which is given to a large variety of chemical changes. In order for something to burn three conditions must be met:

- there must be some fuel – a material which contains large amounts of chemical potential energy;
- there must be some oxygen – that part of the air (about 21%) which is needed for fuel to burn and which humans and other animals need for respiration;

- there must be a high temperature.

Other areas where children can observe and investigate chemical change is through decay of animal and vegetable material and the corrosion of some metals.

Rusting

When impure iron and steel are exposed to the elements they begin to be 'eaten away' – they are corroded. This effect is often referred to as weathering. It occurs when iron combines with oxygen and water to form iron oxide and water.

Children's understanding of chemical changes

As discussed in the previous section, children are familiar with the physical changes to materials brought about by mixing, applying forces and heating. These changes may visibly alter the material, perhaps by changing its state from a solid to a liquid (e.g. when sugar is heated), or by mixing (e.g. when adding salt to warm water), but the materials themselves have not changed; no new materials have been created. However, when some materials cook, burn, decay or corrode then chemical changes take place which do result in the creation of a new material.

Children will have had the opportunity from an early age to observe many examples of chemical changes to materials within a variety of situations. Many of these changes will have taken place in the kitchen, in the context of cooking and food preparation. Foods being cooked may change in texture and colour, particularly if they are burnt! They will also have seen fruit and vegetables left in a bowl or a cupboard ripen and decay, milk begin to sour, and nails, garden tools and bicycle parts show signs of rusting. They may have had the opportunity to mould clay while it is wet and turn it into a shape that will be fixed when the clay is fired. They will accept that these things happen, and they can be encouraged to make careful observations of these changes.

Teaching chemical changes in materials

There are great opportunities to extend and develop an understanding of chemical changes through simple cooking activities (see Figure 3.6 for resources, and the case study in the section on 'Physical changes'). Children can be encouraged to focus on particular aspects of the changes to the materials. What is it about the material that has changed? Is it the colour that has changed? Is it the texture that has altered? Or is it perhaps both, e.g. when bread is burnt to make toast? What conditions have brought about the change? In the case of the toast, is it a change that has occurred simply as a result of an increase in temperature? Bread that has been left out in the air uncovered, can become brittle even though it has not been heated. Consideration can be given to the material that has been produced.

Adding hot water to baking powder, bicarbonate of soda, etc., can provide an opportunity for developing experience of effervescence.

A collection of objects made from different metals could be investigated to see whether all metals weather in the same way. Paper clips, drawing pins, aluminium foil, cutlery, nails, etc., could be examined both before and after exposure to air. A fresh set of the same metals might then be exposed to immersion in water. A further development would be to immerse the metals in other liquids like oil. Do the metals change in the same way? A collection might be buried in the ground and then examined after some time to observe any changes.

The same approach can be taken with other collections, perhaps different types of wood, plastic, seeds, rocks, etc.

MECHANICAL PROPERTIES

Understanding mechanical properties

The properties which govern how materials respond to forces that are applied to them are known as mechanical properties. The properties of any object can be divided into two types, and it is important to understand the differences between them.

A **material property** is a property of the stuff from which the object is made and it is unaffected by the size or the shape of the object. For example, it would be as difficult to push a thumb tack into a 1-cm cube of mahogany as it would into the whole trunk. They have the same degree of hardness. The test that is often used to compare the relative hardness of a material is called the scratch test.

The **object property** of an object depends not only on the material from which it is made, but also its size and shape. For example a thin sliver of mahogany could be bent quite easily, whereas a huge beam would not. This is a measure of stiffness, which is an object property.

The way in which a material will respond to a force being applied to it depends on the bonding between the molecules which make it up. The stronger that bonding is, the more the material will resist changing its shape. Flexible materials such as wood, paper and card, rubber, plastic, threads and fabrics are made up of large molecules which form long, thin chains, which are repeated over and over again. (These materials are called polymers.) The chains can be arranged in lines that all lie in the same direction or they can be tangled up. They can then be bound together sideways onto each other in varying degrees. Large forces are needed to break these chains in tension.

Rubber is weakly elastic because the tangled molecules are only loosely bound to each other. When a rubber band is pulled, the tangled molecules straighten out. When the band is released the molecules tangle up again and the band returns to its original size and shape. This is an example of elasticity.

Cotton thread is not so elastic, because its chains are not tangled up but are arranged in linear fashion. When the thread is pulled the chains cannot untangle, they are simply pulled apart until they break.

It is useful to consider the mechanical properties of materials into six broad areas:

- **Hardness.** This is a measure of how easy or difficult it is to mark the surface of a material with a dent or a scratch. A standard scale known as Mohs scale is generally used. Values are on a scale of one to 10. Talc would be softest at 1 and diamond the hardest at 10.
- **Compressibility.** This is directly related to the structural differences between solids, liquids and gases which has been discussed earlier (see section on 'Structure of materials'). When solids are compressed there is no discernible change to their shape or volume. Because liquids change shape so easily it is rather less obvious to observe that they too are incompressible. Gases, however, are compressible and can be squashed so that the amount of space that they take up is reduced.
- **Elasticity.** Elastic materials are flexible materials which can bend and stretch and will return to their original shape and size when a force is no longer being applied to them, for example a rubber band.
- **Plasticity.** Plastic materials (this does not mean made from plastic) can bend and stretch when forces are applied to them, but do not return to their original shape and size; they remain in their new shape. This is called plastic deformation. An example of a plastic material is modelling clay.

Elasticity and plasticity are material properties which are not mutually exclusive. When describing common objects it may be necessary to refer also to stiffness and flexibility. For example, a large block of rubber and a thin rubber band are both elastic, but the block is much stiffer than the band.

- **Brittleness.** A brittle material will break suddenly, for example glass and pottery. Because a material is brittle it does not mean that it is weak.
- **Toughness.** A tough material breaks slowly, for example wood. This does not mean that the material is strong.

Two areas in which object properties are important are stiffness and strength:

- **Stiffness.** An object which shows no change when forces are applied to it would be termed completely rigid. Degrees of rigidity are frequently referred to in terms of stiffness. Some materials are more resistant to changes in shape through the application of forces than others. The amount of resistance might be termed as the amount of stiffness that the object has.
- **Strength.** Scientifically, strength is a measure of the forces that an object can withstand before it breaks. The greater the force that is withstood, the stronger the object. Strength can be both a material and an object property. It depends on the size and shape of the object as well as the material from which it is made.
- **Compression strength** is a measure of how much squashing force can be applied before the object breaks.
- **Tensile strength** is a measure of how much stretch or pull can be applied and bending strength is a measure of bending before breaking.

Children's understanding of mechanical properties

This is an area where it is most important to ensure the 'correct' use of language, particularly in relation to properties such as strength and flexibility, in order to avoid confusion. This is more likely to occur when children confuse hardness, toughness and brittleness with strength.

It may be useful, if appropriate, to involve the children in the construction of a concept map in order to gain some insight into their understanding of these terms and ideas. Concept maps have been discussed in Chapter 1 and there are examples at the end of this chapter (Figure 3.7) and in Chapter 4.

Difficulties can develop where there is a developing level of understanding from other areas of study. For example, it can be difficult for young children to develop a clear understanding of compressibility and non-compressibility until they have an understanding of conservation of volume. They may also benefit from linked experiences with materials that are plastic and elastic. Given a piece of Plasticine to squash, the child may see that the material can be compressed simply because it has changed its shape. The teacher might ask the question, 'Have we made the Plasticine smaller, does it take up less space now?' Questions about whether or not the Plasticine will return to its original shape will emphasise the plastic element.

There is an opportunity here to investigate these materials by relating the changes, either reversible or not, to the amount of force that has been applied. The rubber band will stretch more as more force is applied to it. If a Plasticine ball is dropped from a certain height and the changes to its shape are observed and then another identical ball is dropped from twice the height, an increase in distortion is observed.

Children will often confuse stiffness with strength. If a material is easily bent, like a thin copper wire, children will often say that the wire is weak. If they were able to apply the same force to a thick copper rod they would say that its resistance to the force, its unwillingness to bend, shows its strength. This thinking reflects the confusion between strength and the wire's stiffness.

Young children also view solids as strong, hard and rigid. It is therefore important to provide them with as wide a variety of experiences as possible to enable them to develop an understanding that solid materials can be soft and non-rigid and will include powders, which are often considered to be liquids. Fabrics are often thought of as being somewhere in between.

Teaching mechanical properties of materials

There is enormous scope here for the exploration and investigation of a rich variety of materials. Virtually any available material can be used as long as health and safety issues are considered. Curiosity can be fostered through the use of surprising materials like 'Silly Putty' and cornflour mixed with water.

Young children can compare the hardness of one material with another by simply trying to dent or scratch one with another. With younger children, the number and range of materials available for investigation should be limited so that a simple hierarchy of hardness can be established. Materials might include clay, balsa wood, a steel nail. The clay can be dented by each of the others and so it must be the softest. The nail will dent the balsa wood, etc. The investigation could be developed through extending the range, looking at softer than and harder than materials.

The range of materials can then be extended to include perhaps ranges within one type of material, for example a range of woods (including balsa, pine, mahogany and, if possible, very hard woods like ebony) and a range of rocks including talc, chalk and granite.

Care must be taken in order to distinguish between the scratching of one material and the rubbing off from one material to another. A nail will not be able to scratch the surface of a piece of granite but some of the metal may be rubbed onto the rock.

The importance of applying the terms 'solid', 'liquid' and 'gas' to the substances making up materials rather than to the materials themselves may be seen here. A sponge, for example, would be described as a solid. However, it is a mixture of substances. The air between the sponge allows the substance to be compressed.

Children need to relate the characteristic properties of solids (definite shape and non-compressibility) to the substances of which powders and fabrics are made by looking at individual grains and fibres. Using magnifiers the children can be encouraged to see that a grain of sand has its own shape.

ORIGINS, MANUFACTURE AND USES

Understanding origins, manufacture and uses

Materials are found in the Earth's atmosphere, in its seas, in its rocks and in the plants and animals that live on it. Materials are everywhere: oxygen and ozone, so important for our existence, are found in the atmosphere; salt is found in the oceans; wood is found in trees; and diamonds are found in rocks. When investigating materials, it is useful to consider whether the material is synthetic or whether it is naturally occurring. These distinctions can be made from the earliest experiences.

Natural materials are produced by natural processes and changes. These include animal and plant materials such as bone, hair, wool, wood, silk, and cotton, as well as mineral materials such as granite, salt and gold.

Natural materials can be changed physically by humans in many ways in order to make them more suitable for a particular purpose. This does not change the nature of the material itself. For example, slate can be cut into shape so that it can more easily be fitted to form a roof. The cut slate is a man-made natural material. There are many changes which the children may be familiar with. These may include:

- spinning, weaving, knitting and plaiting; using wool, linen, silk, cotton and paper;
- carving and cutting; using wood, clay and stone;
- sieving; using gravel, coal and flour;

- dissolving, filtering and evaporating; using salt and sugar;
- crushing and grinding; using cereals, chalk, spices, rock and metal ore.

Synthetic or manufactured materials are made by processes, which are often chemical, changing a raw material into a different material. Here the nature of the materials have changed. The use of crude oil to manufacture plastics, artificial fibres and many other materials is probably one of the best examples of these types of change.

It is also of use to have an appreciation of the properties and origins of materials in order to understand why particular materials are used for particular purposes. Some of those materials which are used most commonly in the children's experience are mentioned below.

- **Wood.** There are two main groups of wood. Those produced by deciduous trees such as oak, which are hard woods, and those produced by coniferous trees such as pine. Woods vary in hardness and this often determines their use, although compared to materials such as stone they are relatively soft. Woods are strong and stiff under compression and tension and yet elastic when bent. These properties mean that wood is a very useful material. Its strength means that it is used in house building for roofs and load-bearing joists. Hardwoods are also used for doors and some exposed flooring where durability and a good appearance are required.
- **Glass.** This man-made material is made from two natural materials, sand and limestone, plus soda-ash, which is a manufactured chemical. These are heated together until they melt. This liquid can be poured and drawn out into sheets while it is still liquid. Once cooled and a solid it is easily cut into the required shape and size. Glass is transparent, hard and smooth, which makes it ideal for windows, strong and stiff in compression and tension, yet flexible and elastic when bent. It is very brittle, which is a disadvantage, although this can, to some extent be overcome by combining thin metal strips with the glass to reinforce it, allowing greater flexibility.
- **Plastic.** Almost all plastics are made from chemicals which have been obtained from oil. Some plastics are flexible and tough and are used to make pipes and guttering, covering for electric cables and polythene sheeting. Other plastics are harder and are used to make plugs and sockets as well as kitchen work surfaces. Most of our food packaging is made from plastic, as it is easy to mould and relatively light.
- **Clay.** Clay is mainly composed of aluminium, silicon and oxygen. These are arranged in separate layers which can slide over one another when wet. This is why clay can be moulded so easily. When the clay is heated in a kiln, the water is driven out and the layers are bonded together to form a rigid structure.
- **Metal.** Most metals are obtained by a process called smelting, which extracts the minerals from ore-bearing rock. Sometimes mixtures or alloys are made in order to alter the properties of the metals. For example, copper and tin are mixed to produce bronze. The higher the proportion of tin to copper the harder the resultant bronze will be. Brass is an alloy of copper and zinc and steel is an alloy

of iron and carbon. Metals vary in hardness, and for their size and weight are very strong and stiff in compression and tension. Metals are tough and easy to shape and mould. Depending on how it is shaped, metals can be flexible (e.g. in a wire) or rigid (e.g. in a girder). The main disadvantage of some metals, particularly iron, is that they corrode easily. Other metals like gold, platinum and silver do not.

Children's understanding of materials' origins, manufacture and uses

Many children are prepared to accept things as they are, until there is a need to question something. They believe that objects and the materials that they are made of have always existed as they appear. Young children will consider that 'things come from shops'. Older children may begin to think in terms of 'things being made in factories'.

It is not until children are aged about 7 or 8 that they begin to accept and recognise that some materials have come from plants or animals and been changed in some way. The idea that metals come from particular rocks, that plastic is made from oil and glass from sand are even more difficult to comprehend because it is unlikely that the children have had experience of the manufacturing process.

Young children also find it difficult to recognise living things as being made of material. That bones, teeth and hair and other parts of their bodies are also materials is particularly curious for them.

Children have much less difficulty in considering the uses of materials. Early work in the nursery, particularly with 'junk modelling' encourages the selection of the 'right thing for the job'. This is developed throughout the school in many curricula areas.

Teaching origins, manufacture and uses

The clearest way of helping children to have experiences of the manufacturing aspects of materials is to enable them to extract and make materials for themselves.

- Samples of sea water can be collected (or if this is difficult, created using salt and warm water) and small quantities allowed to stand so that the water evaporates leaving the salt.
- Samples of wheat and other cereals could be collected and then ground and sieved.
- Paper could be collected and made into pulp.
- Wool samples could be combed out and spun or weaved. Natural dyes made from onion skins or berries could be used to colour the wool. Be aware of Health and Safety Regulations concerning the collection and handling of berries.

Visits to factories, industrial museums, science museums and farms are also very useful. Experience can be supplemented through the use of secondary sources such as videos, films, books, magazines and CD ROMs.

For children to understand the uses that particular materials have, they will need to have experiences which will further their knowledge of the properties of materials. By carefully selecting the choice of materials available we can focus on particular properties of materials such as transparency, flexibility, solubility and others, some of which are discussed earlier. A starting point may be to look at some of the materials that make up the objects in the classroom.

- What are the classroom chairs made from?
- Why do you think that they are made from wood or plastic?
- Could we have chairs that are made from another material, like glass?
- Why are the windows made from glass?
- Could the windows be made from any other material?

The last question leads the children towards thinking about the common properties of some materials. Older children may be given some problem solving situations to consider. For example, I need to transport a very delicate glass bowl to Australia. What are the best materials to use to protect it? Each material selected could be fair-tested to ascertain its suitability.

An awareness on the part of the teacher of the many opportunities that there are to develop and extend the child's understanding of the properties of materials and their uses, through a breadth and depth of experience both inside and outside the classroom, will result in valuable scientific investigation.

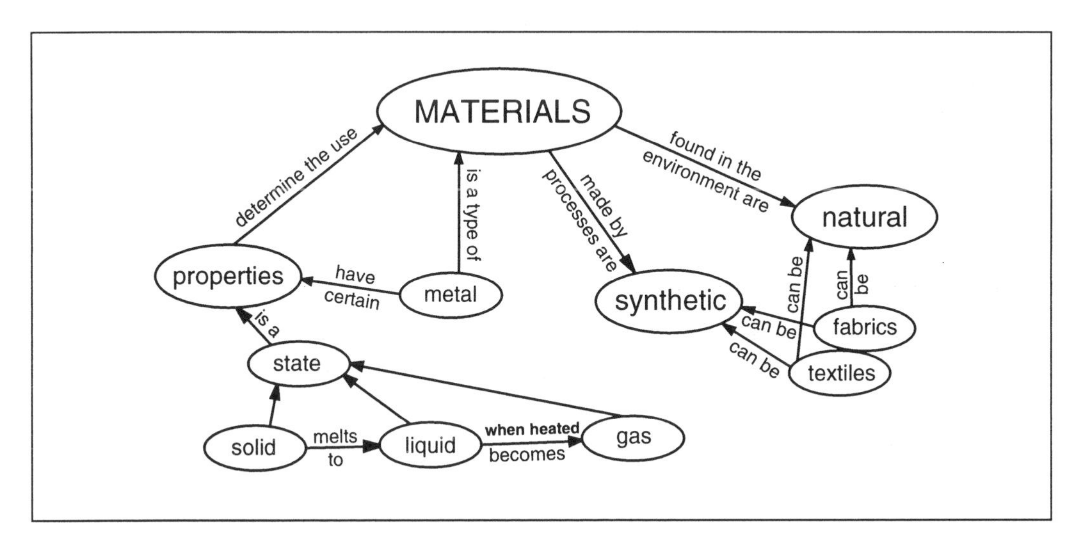

Figure 3.7 Student's concept map

At the end of this chapter you might like to try to draw a concept map to summarise your own understanding of the ideas presented. Figure 3.7 shows a student teacher's attempt to cover some of the areas. Do you agree with these relationships? How does your map compare?

Looking at the National Curriculum for materials as a whole, it is possible to identify progression in the ideas, by reference to the level descriptors; Figure 3.8 shows a summary of these, for reference.

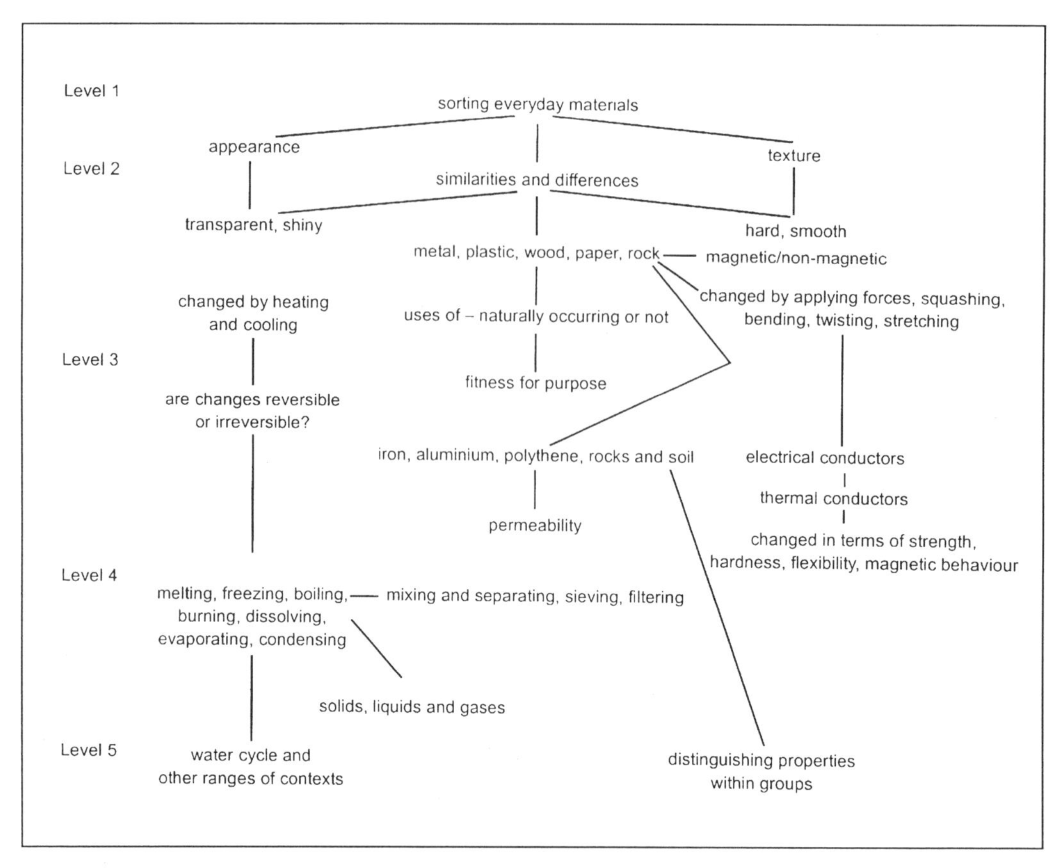

Figure 3.8 Progression of ideas in the National Curriculum

CHAPTER 4

Physical Processes

ELECTRICITY

Understanding electricity

What is electricity?

Electricity is a flow of **electrons**. This flow is called an **electric current**. This can produce many effects such as lighting a lamp and sounding a buzzer. These all depend on the transfer of **energy**.

This summary shows why there is often a confusion for children (and adults) about electricity. Electrons and their flow cannot be perceived directly, only their effects, so it is often these effects that are referred to as electricity. In this section we will first consider electrons, then their flow in circuits as current, and finally the effects of this as energy transfers.

As was described in the previous chapter, all substances are made of atoms. All atoms consist of a central nucleus which has a positive charge, surrounded by electrons which have a negative charge. Electrons can be removed from or added to atoms, which then leaves them with a positive charge, or an extra negative charge. Charged atoms are called **ions**. With some materials such as plastics, charge can be transferred just by rubbing with a cloth, e.g. a rubbed comb will attract hairs or small pieces of paper (Figure 4.1), and synthetic underwear sometimes crackles when taken off.

These effects show how charged objects can be recognised: there are forces between them. A charged object attracts other objects. Careful investigation shows that oppositely charged (i.e. positive and negative) objects attract, and similarly charged objects repel (Figure 4.2a). Note the rule is similar to that for magnetic poles (Figure 4.2b), but there is no connection between the two effects.

The crackling or sparking of charged objects is called **discharge** and is caused by the charge flowing away. Materials which can hold charge are called **insulators**, those which allow it to flow are **conductors**. Plastics and rubber are insulators, metals are conductors. The phenomenon of charged objects, **electrostatics**, was known at the time of ancient Greece; the word electron comes from the Greek for amber, a material easy to charge by rubbing. By contrast, the association of this with current is relatively recent and results from the work of Benjamin Franklin on lightning (mid-eighteenth century), and of Alessandro Volta on the battery (late eighteenth century).

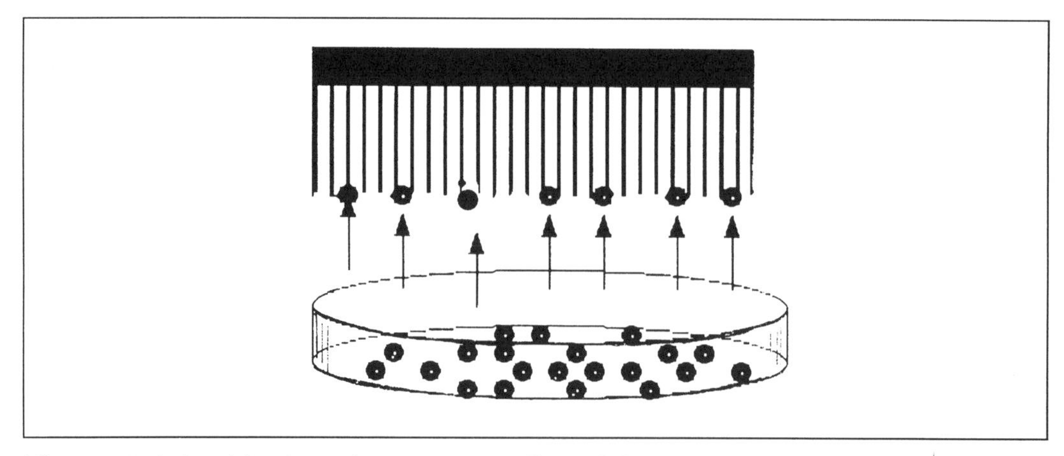

Figure 4.1 A rubbed comb attracts small particles

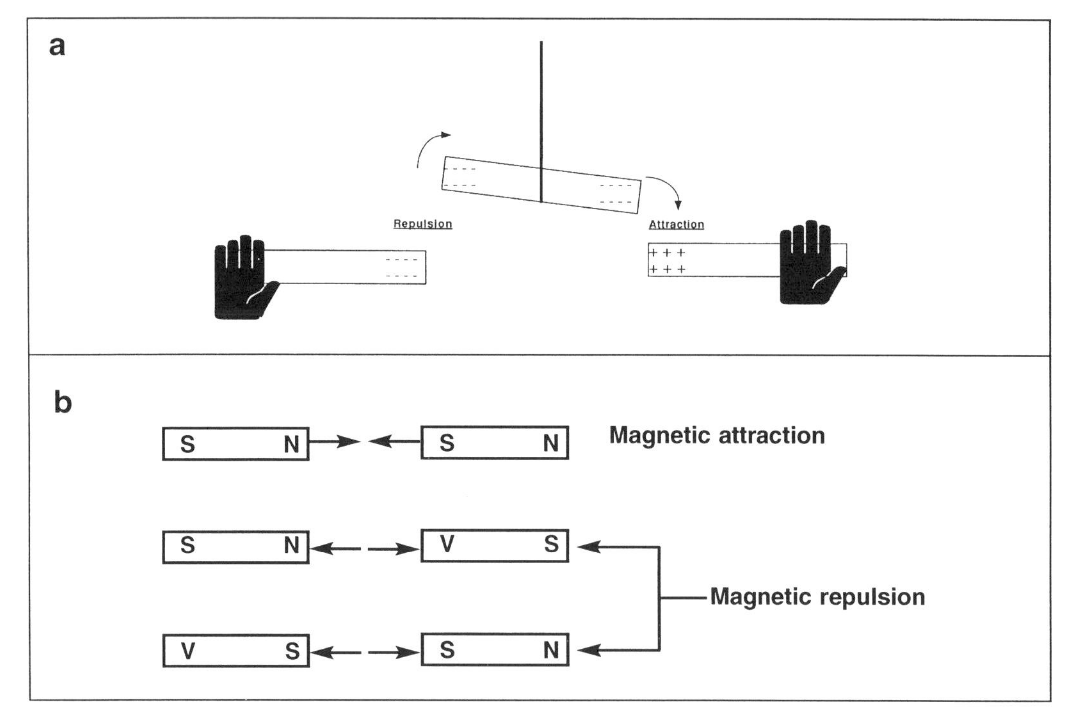

Figure 4.2a, b Like charges repel, opposite charges attract; like poles repel, opposite poles attract

Circuits and currents

For a current to flow there must be a source of energy, such as a cell or battery (several cells connected together), and a conducting route. This route must go from one terminal of the cell (marked positive) to the other (marked negative). The current can travel through components such as bulbs and buzzers, if there are suitable connections, i.e. an 'in' and an 'out'. This is called an electric **circuit**, and no current will flow unless it is complete. The flow can be interrupted with a switch, and affected by the components in a circuit. Circuit components and arrangements can be shown by the use of symbols (Figure 4.3).

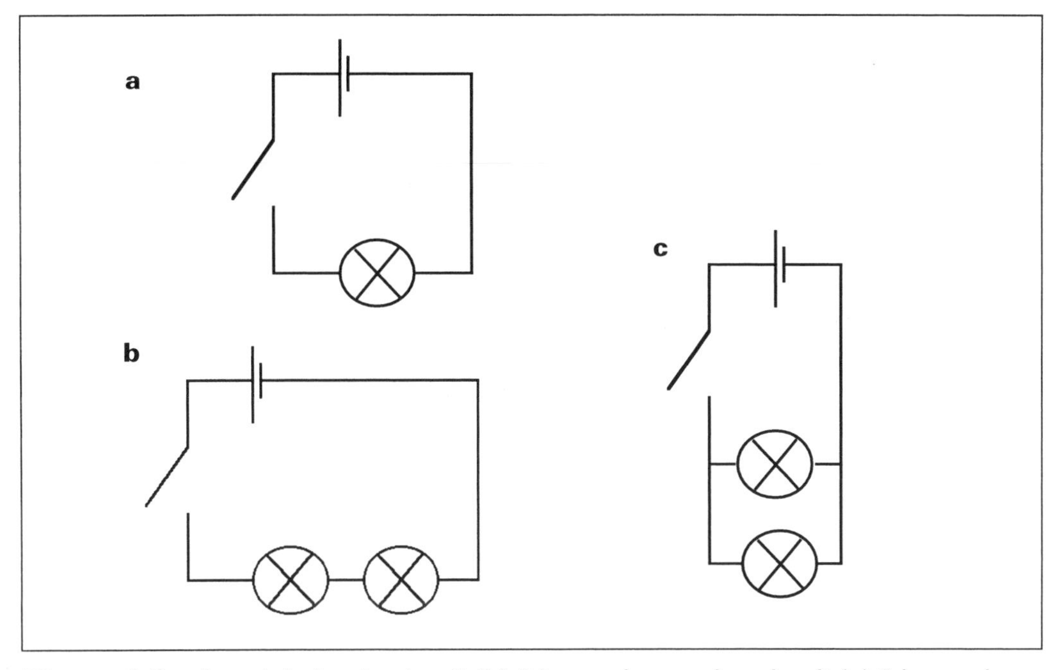

Figure 4.3a, b, c (a) simple circuit (b) 2 lamps in a series circuit (c) 2 lamps in a parallel circuit

The two lamps in the circuit in Figure 4.3b will be less bright than the lamp in the circuit in Figure 4.3a. This arrangement is called a **series circuit** because the current goes through all the components in turn. The arrangement in Figure 4.3c shows two lamps in **parallel**. Here the current has two alternative routes. Part will pass through one lamp and part through the other. Each route is equivalent to the first figure with only one lamp (check this by tracing the routes), so each lamp lights as brightly as the first one.

The lamps light because the current causes them to heat up. Energy is transferred to the lamp by the current. This is because the thin wire in the lamp, the filament, has **resistance**. The energy is transferred from the cell, where it is stored in the chemicals. These react when a circuit is connected and the electrons are driven round by forces of attraction and repulsion, like those between charged objects.

A cell has a certain **voltage (V)** rating. The commonest cells are 1.5V, and these can be joined to make 3V, 6V and 9V batteries. It is the voltage of the cell which measures its **potential** for driving electrons round a circuit. The combination of the voltage of the battery and the resistance of the lamp(s) will determine how much current flows in a circuit, and therefore how brightly the lamp lights. The current is measured in amperes or amps (A) with an ammeter; resistance is measured in ohms (Ω) with an ohmmeter. The symbols for these are Ⓐ and Ⓘ, and that for a voltmeter is Ⓥ. **Multimeters** are useful for home or school use because they can measure all three quantities. There is a dial and a choice of settings for each scale (Figure 4.4).

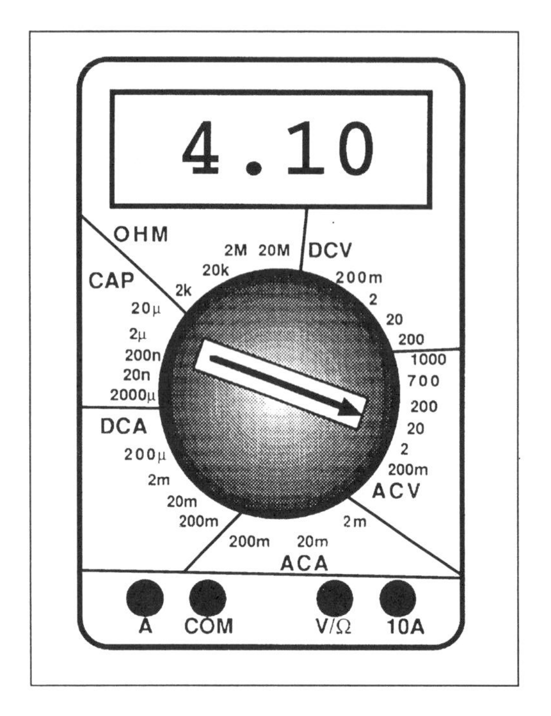

Figure 4.4 Multimeter

To measure current, the circuit must be broken and the meter inserted so the current flows through it. Measurements show that the current is the same all round a series circuit, but splits and then rejoins in a parallel circuit (Figure 4.5).

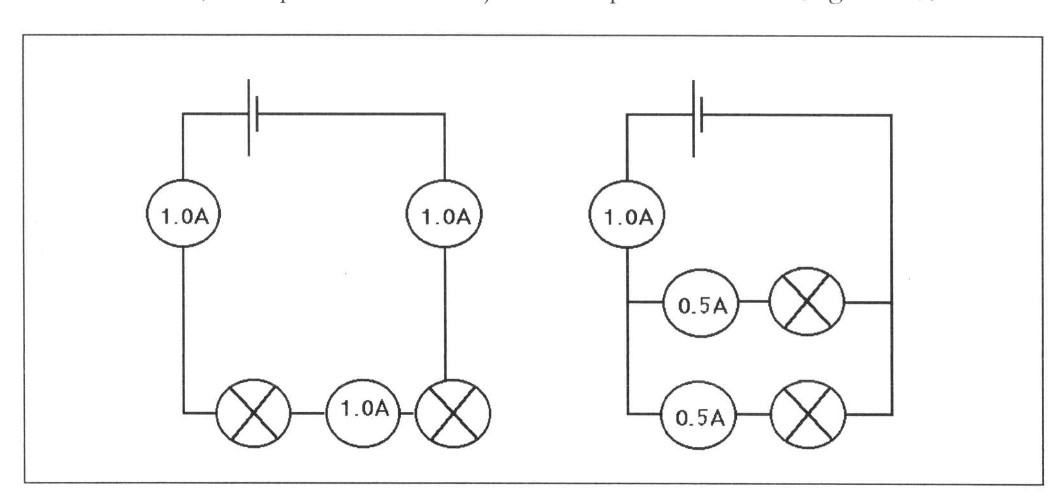

Figure 4.5 Measuring current in a series circuit; measuring currents in a parallel circuit

To measure voltage, the meter must be connected across the cell or other component, without breaking the circuit (Figure 4.6). Measurements show that the voltage is provided by the cell or battery and 'used up' by the bulb or other component.

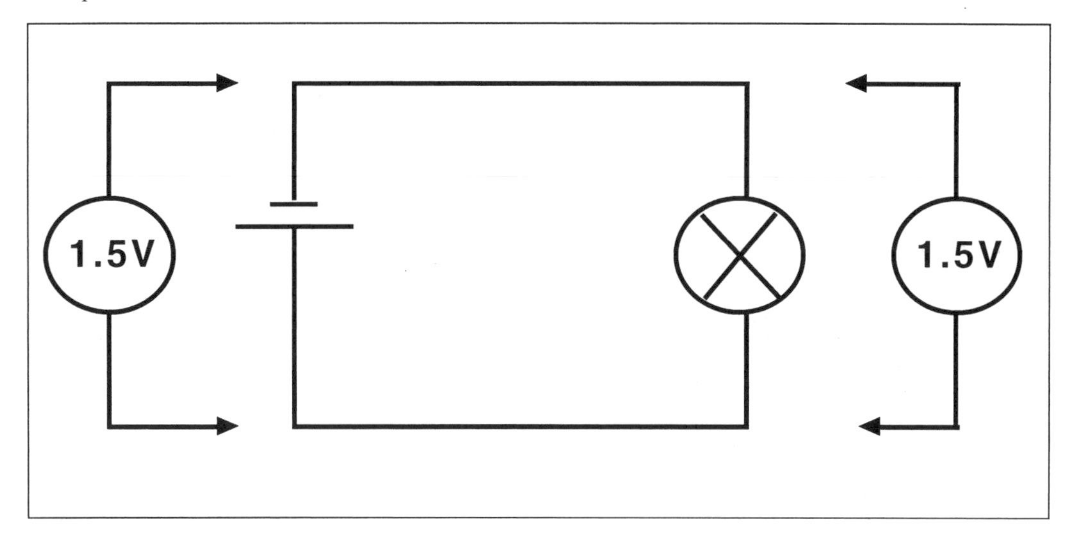

Figure 4.6 Measuring voltage in a circuit

A circuit is not needed to measure resistance; just connect the component to the meter. This is a good way to test if a component is broken, e.g. if a bulb or fuse has broken, its resistance will be infinite. The relationship between the resistance of a component, the voltage across it and the current through it is given by **Ohm's Law**, which says:

Current [A] = voltage [V]/resistance [Ω]

So the current in a circuit can be predicted, e.g. if the voltage is high or the resistance is low, there will be a high current.

Making and using electricity

A cell can be said to 'make' electricity in the sense that it is needed to make a circuit work. What this means is that it provides a voltage (because of the chemical change in it) to make the electrons move. So energy is carried in the circuit to do useful things like light a lamp. Most of the electricity we use is 'made' at power-stations, and supplied to us at mains sockets.

Power-stations are large factories for transforming energy. The source of energy is usually a fuel (coal, gas, oil or nuclear). This is used to produce steam under pressure which drives a turbine and a **generator** to make electricity (Figure 4.7). Power-stations are often very big so they can generate a large electricity supply economically.

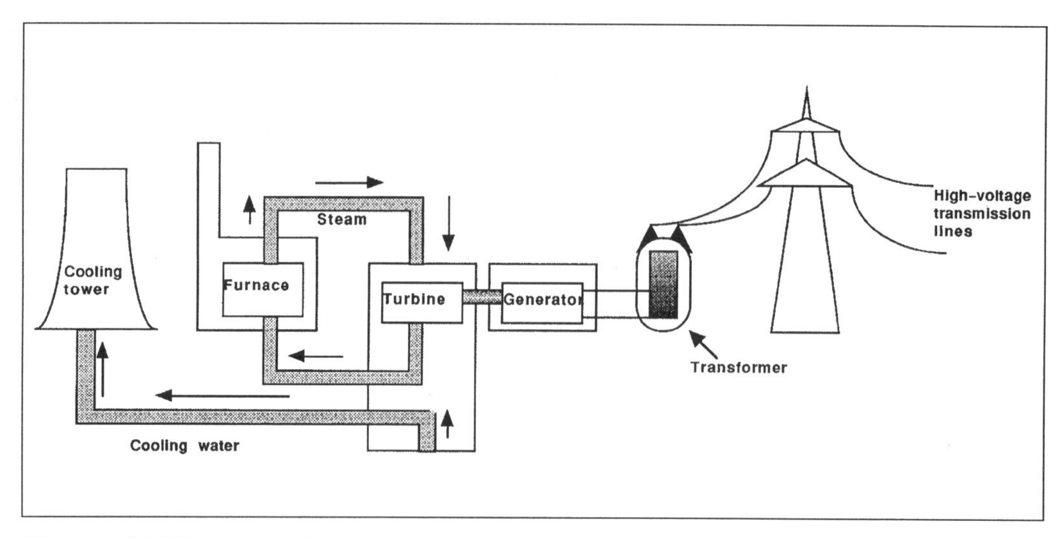

Figure 4.7 Power-station

The principle of the electric generator was discovered by Michael Faraday in 1831, and it depends on the fact that a changing magnetic field can produce a voltage across a conductor, so that a current will flow. Power-station generators are larger versions of the dynamos used to power cycle lamps. Each consists of magnets and coils of copper wire which are free to rotate (Figure 4.8).

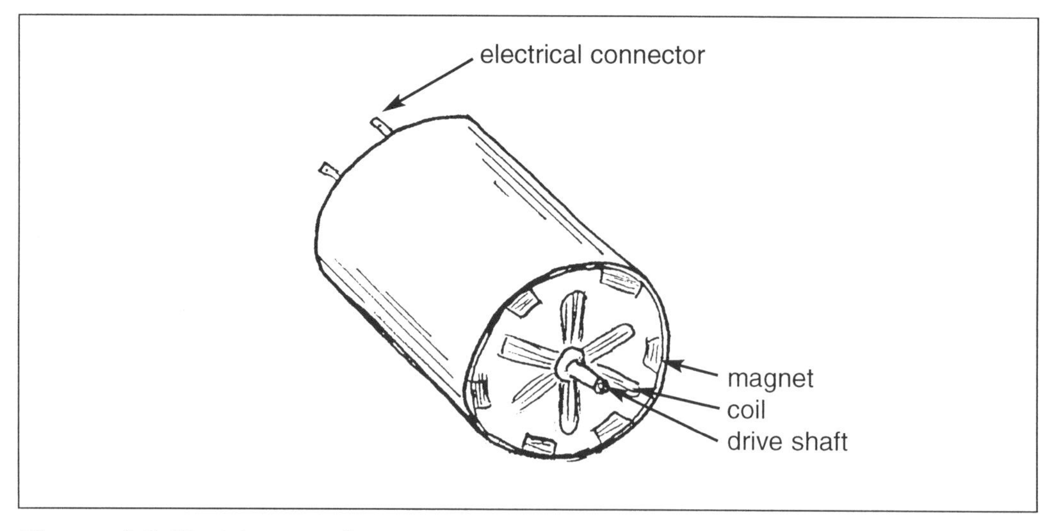

Figure 4.8 Electric motor/generator

Electricity is generated in a power-station at a voltage of about 25,000V. Because of the rotation of the generator, the voltage reverses 50 times a second, so the current direction alternates. This is called a.c., alternating current. It has the advantage that it can have the voltage changed by **transformers**. At the power-station the voltage is transformed up to about 400,000V and carried around the country on a network of cables called the National Grid. High-voltage transmission reduces the energy loss, but the cables have to be carefully insulated to prevent leakage, usually by hanging them from tall pylons. The voltage is then transformed down for use in industry (11,000 or 33,000V) and to the home (230V – 240V before January 1995). All these voltages are hazardous (see later).

We use electricity so widely because it is a convenient, clean and relatively safe way of transferring energy. Energy is needed to make things work (see next section). Electricity transfers energy by the movement of electrons. The appliances convert this into other forms of energy. In lamps and heaters, resistance to electron flow causes an increase in temperature and the radiation of light and heat. In audio equipment such as a loudspeaker, the current is used to make an electromagnet. This is placed near a magnet which causes a vibration to generate sound. An electric motor, used in all electrical machines, also works on the principle of the forces between an electromagnetic coil and a magnet. In this case the current causes a rotating motion. It is the reverse of the generator effect. The same device (Figure 4.8) can generate electricity from motion or motion from electricity.

Electrical hazards

Figure 4.9 Electric hazard sign

Electricity presents two hazards to health. As an energy transfer method it can cause burns to the body and fires. As a flow of electrons it can cause shocks by interfering with the electrical signals of the nervous system. The danger depends on the current, but as Ohm's law shows, this is determined by the voltage of the supply and the resistance of the circuit. We rarely know the resistance of a circuit, so it is safer to have rules based on voltages.

Batteries have voltages up to 12V and are inherently safe from giving shocks, though large capacity batteries, such as those in cars, can produce heating effects which could burn. Also there can be chemical hazards from the acid they contain, so they should not be used in primary schools. Rechargeable cells can get hot if discharged quickly, e.g. by a short circuit. They should only be recharged by adults using a charger designed for the purpose; again the batteries can get hot.

Mains electricity is at 230V and can cause fatal shocks, as well as burns and fires. Experimenting with the mains is forbidden, and only approved mains equipment may be used in schools. Regular checks should be made to ensure there is no damage to this equipment or to the plugs and sockets used. Approved appliances and circuits have two protection features. First, the earth wire provides a connection between metal parts of the appliance and the ground, so that if a faulty live wire should connect to this then current flows to earth, not through the user's body. Second, the fuse protects against current overload. If there is a short circuit due to a fault, too much current could damage the appliance or cause a fire. The fuse melts instead to cut off the current. Circuit breakers are now often fitted into mains supplies; they cut the overload current by an automatic electromagnetic switch.

High-voltage supplies such as the National Grid and railway power lines are exceptionally dangerous, and children need to be warned to keep well clear as sparks can carry a high current to a body even if it is not in contact with a conductor. By contrast, the voltages in electronics and microelectronics circuits such as computers are very small. Electronic kits and logic circuits are powered by low-voltage batteries. Components are often made of partly conducting materials called semiconductors. They can easily be damaged by excess voltage and need to be protected from sparks due to the build-up of static.

Electrical concepts and quantities

One of the confusing things about electricity is that there are several interconnected quantities, each of which can be detected by their effects, not directly. One way of dealing with this is to use a model or analogy which connects

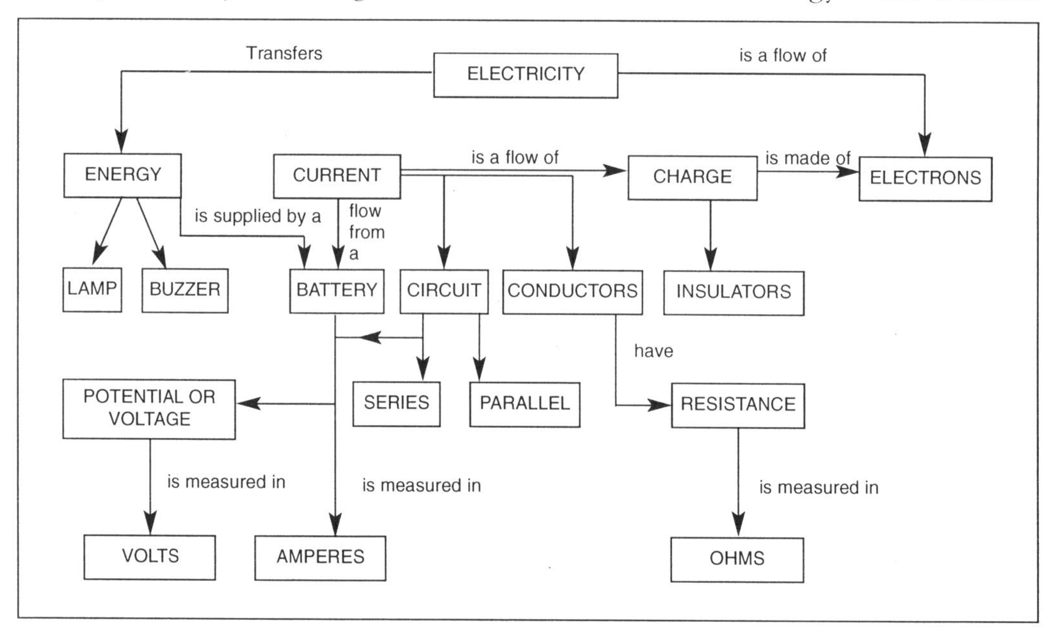

Figure 4.10 Concept map for electricity

these to things that can be more directly experienced. For example, current can be thought of as similar to a flow of water. The use of models in teaching electricity is discussed in the next section.

Another way of checking our own, or the children's, understanding is by the use of a concept map, specifying relationships between all the quantities. Figure 4.10 shows some of these relationships for the concepts presented in this section. You could copy it out and add further links to check your understanding.

Children and electricity

Knowledge and understanding

Teachers, aware of the conceptual complexity and practical hazards of electricity, are often apprehensive of introducing this topic into the classroom. However, even very young children can build and control simple battery-operated circuits in the classroom, and they operate electrical appliances in everyday life without any awareness of this. They approach problems at the practical level of getting things to work. The challenge for the teacher is to develop this natural curiosity and drive for technical competence, into scientific understanding. At primary level the conceptual framework need only be quite modest. Research from the SPACE and CLIS projects has shown many children often have the following ideas:

- **Sources of electricity.** Electricity is often thought of as a kind of substance which comes from the battery and as the battery runs down it has lost this substance – so children expect it to be lighter. Electricity is a material which goes to light the bulb. Young children will not have a particulate concept of matter and so will be unable to appreciate the distinction between an electron and the energy it transfers. The challenge is to develop the idea of the battery as providing the non-material quantity energy.
- **Current in a circuit.** Children may adopt the idea of a flow of something from battery to bulb; Figure 4.11 shows three possibilities. Exploration with several bulbs in different arrangements can help children develop the idea that in a simple series circuit the current is the same all the way round (e.g. all the bulbs have equal brightness). They also need to observe that the length or shape of the connecting wire doesn't affect this, and that a switch affects all of the circuit simultaneously. They can be encouraged to develop and use models (see later). Routes starting and ending at the battery can be traced in branching circuits and parallel arrangements to reinforce the need for a circuit. Problem solving with faulty circuits is a useful consolidation activity.
- **Energy transfer by electricity.** The effects of electricity can be experienced by children at an early age. Wiring up model houses to light them and provide bells can be successful in Year 1 (Figure 4.12). The eventual challenge will be for them to differentiate the material nature of the current (not used up) from the energy transferred to lamp or bell. This for most children will be in late primary or early secondary school.

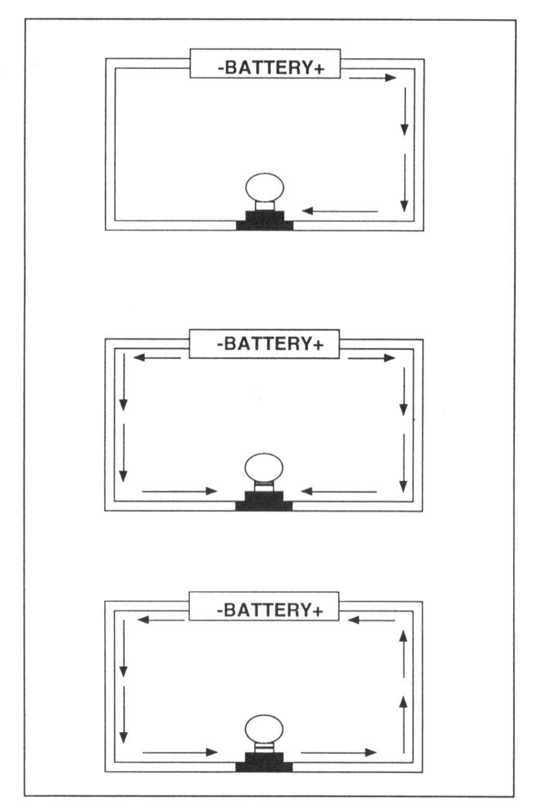

Figure 4.11 Models of electricity flow

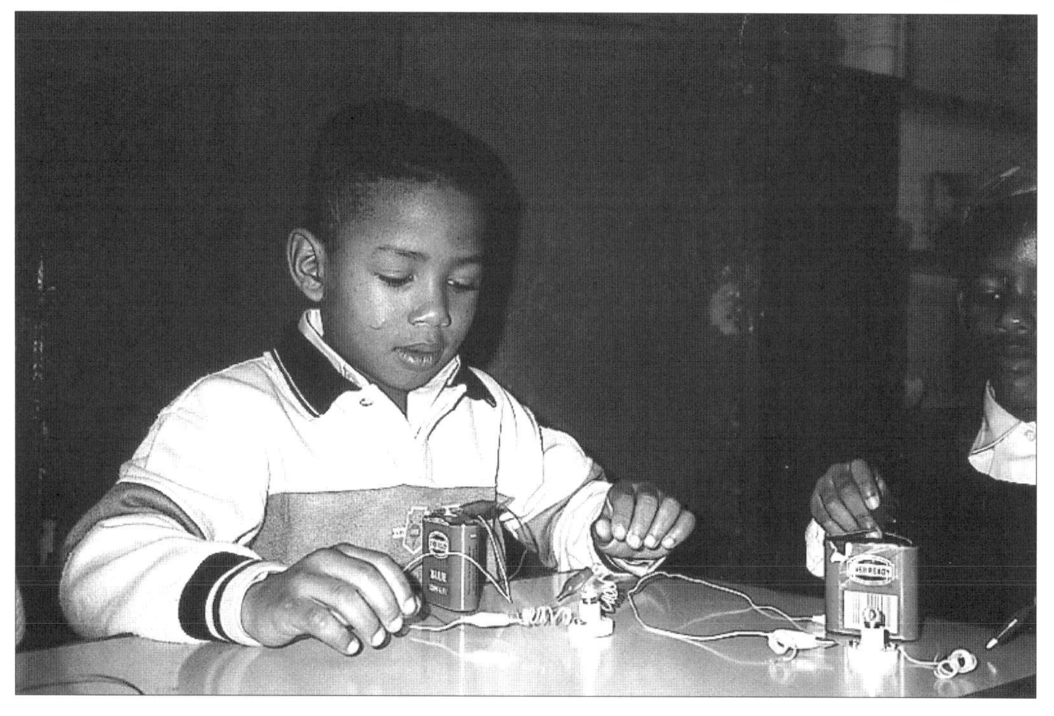

Figure 4.12 Young children undertaking wiring

Skills and attitudes

Children can respond well to the practical opportunities of circuit building at an early age, providing they have a fair degree of success in making things work (see later). Achievement through technical competence can be very motivating. Links to everyday home life and later to the worlds of work and leisure will provide appropriate contexts and develop positive attitudes and a critical approach. Careful observations and pattern seeking and interpreting are key skills rather than those of elaborate 'fair testing'. For older children the need to record circuits with symbols, and to interpret these to build circuits, is a good communication skill to develop.

Teaching electricity

Planning for progression

Considering the topic of electricity in the primary school as a whole, it could be covered in three stages, each time reviewing and extending the experiences and ideas of the previous treatment of the topic:

- **Key Stage 1** (Years 1 or 2). Use of everyday electrical appliances, what they do and how to use them safely; wiring circuits with a bulb and a battery, perhaps to illuminate a model; exploring magnets and magnetic materials.
- **Lower Key Stage 2** (Years 3 or 4). Dangers of electricity – shock and fire; useful circuits including switches and bells or buzzers – how to check and repair faults.
- **Upper Key Stage 2** (Years 5 and 6). Dangers of mains and high voltage power supplies; controlling circuits including series and parallel connections; circuit symbols; electromagnets.

Organising practical work

Exploration of different circuits doing different things is excellent for small group work. All practical work in primary schools will use cells and batteries at low voltage and so is inherently safe.

A suitable equipment list for each group would include: cells, bulbs in holders, switch, dimmer, buzzer and perhaps a motor–generator, copper wire and a large iron nail to make an electromagnet. Full details are provided in Appendix 3 Equipment.

There are often problems with equipment not working. Here is a checklist to help with the identification of faults:

- Is there a complete circuit?
- Are all the connections secure?
- Has the cell or battery run down through use?
- Has the lamp or other component 'blown' – broken or fused because of the current?
- Are the voltages matched? Check that the rating on the component is similar to the voltage of the battery.
- Is the component connected the right way round? Some buzzers and other

electronic items have to have one terminal connected to positive and the other to negative, i.e. electricity can only flow one way through it.

Teach older children to use this or a similar checklist so that they can explore circuits freely.

Safety

- Children need to know from the start that the kind of investigating they do with battery-operated circuits cannot be done with mains electricity.
- Work on how we all use electricity in everyday life can provide the context for devising rules for the safe use of appliances.
- Older children need to be cautioned about very high-voltage supplies such as substations (transformers) and rail power lines. Useful videos are listed in the Appendix.

Concepts and models

With older children it is useful to introduce a simpler version of the concept map (Figure 4.13) to help them check their understanding.

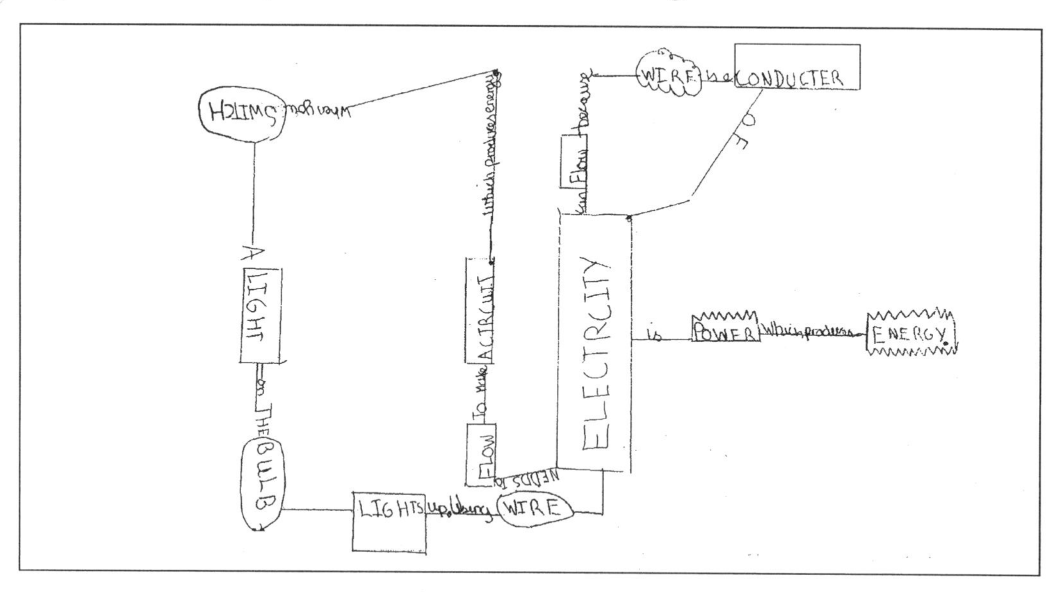

Figure 4.13 Child's concept map

At various ages models can be used to help make ideas more concrete, for example:

- current represented by water flow, so the battery becomes a pump and the lamp a constriction to the flow, or an uphill gradient;
- flow in a circuit is represented by a loop of string: all children hold it in one hand; one child as battery pulls it round, another as lamp squeezes while it passes, and a third as switch breaks the string;

- (more advanced) children as particles of electricity travel around the circuit of other children, are given a sweet to represent energy at the battery and give up the sweet as they pass the lamp.

Developing knowledge and understanding

Here are some quotes from children on key areas in electricity with their ages in parentheses:

1. Sources and uses of electricity
   - 'electricity comes from a battery; a torch has two so that one can be a spare' (8);
   - 'electricity comes from underground' (5);
   - 'electricity keeps us warm and gives us light' (6).
2. Electric circuits and components
   - 'there is a wire from the battery to the lamp, so it should light' (8);
   - 'this is a circuit and it works because it hasn't got any gaps in it' (8);
   - 'a circuit works like a Scalectric track' (9);
   - 'the dimmer works 'cos the electricity can't get through so well' (9);
   - 'the fuse burns up so the electricity stops' (9).

Talking to children about their ideas before and during activities in electricity is helpful to find out where children are in their thinking. Refer to previous parts of this section to decide what activities or discussions would help the children who made these comments to progress in their understanding.

FORCES AND ENERGY

Understanding forces and energy

Forces and energy are often thought of as difficult areas for both children and adults. The concepts are intangible, yet seem to pop up in all topics in science – that is what makes them important. Lay language often does not clearly distinguish ideas that need to be sorted out in science – to understand the connections between force, movement and energy, for example. In this section we will first consider force and its relationship to movement, then energy, and finally show how the two concepts are related.

What are forces?

The simple scientific description of a force is a push or a pull. This corresponds with our everyday, personal experience of physical exertion – what we feel with our bodies. We might push a car, or lift (pull) a box. Inanimate things can also exert forces, of course. If we let go of the box, it will fall because of the pull of gravity. If the car is on muddy ground it will be harder to push because the mud is sticky and pushes back on the car with a force of **friction**. If we drop a sheet of newspaper it will waft slowly to the ground because of the force of **air resistance**. The newspaper can float on water because of the **upthrust** of the water. A balloon will float upward if filled with helium because it is lighter than air, so the upthrust of the air pushes it up. If you rub a balloon on your hair or jumper it will stick to the wall because of **electrostatic force**; the balloon has become charged. Some kinds of materials are affected by **magnetic forces** which make them pull together (attract) or push apart (repel).

All these different types of forces can be considered in the same way. Forces always act in a certain direction: e.g. on Earth gravity always acts downwards, so it makes things fall; friction and air resistance always act in a direction opposite to movement, so they slow things down.

We can represent forces in diagrams (Figures 4.14, 4.15, 4.16) with arrows which show:

- the place where the force acts on the object;
- the direction of the force;
- the size of the force, by using a scale.

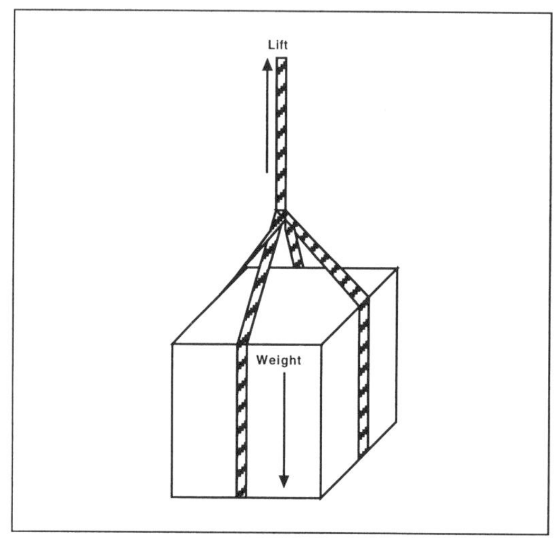

Figure 4.14 Lifting a box

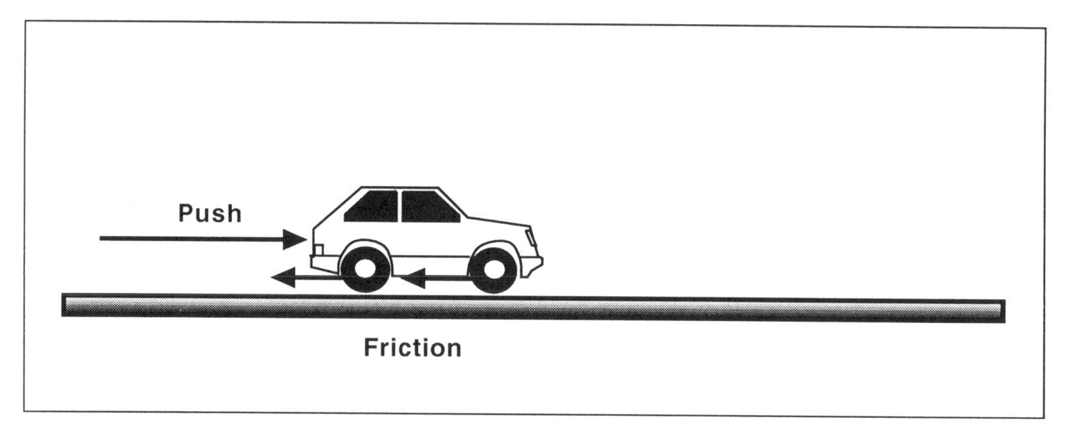

Figure 4.15 Pushing a car

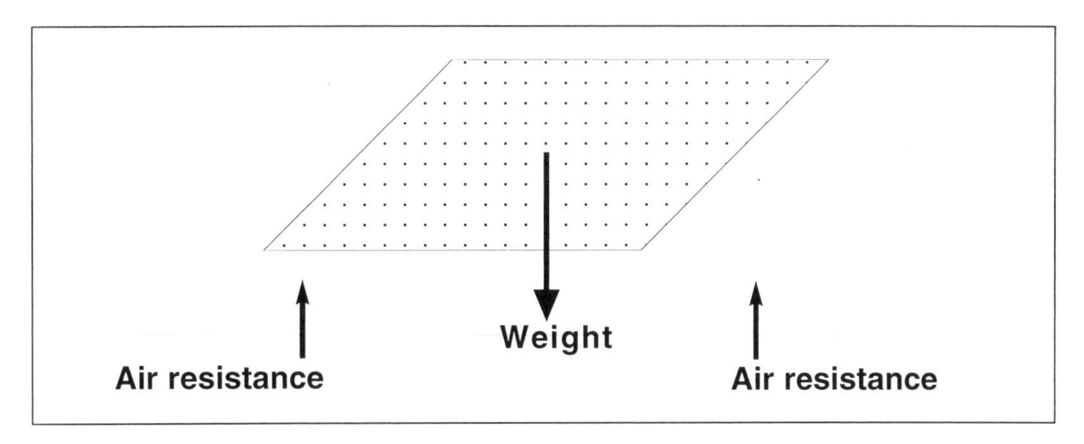

Figure 4.16 Falling paper

The arrows can be drawn to scale to show the size of a force, because force is a quantity that can be measured. A force meter is usually made from a spring because the spring stretches in proportion to the size of the force.

The unit of force is the newton (N). Like any other unit, such as the metre, it is chosen for convenience. Its name commemorates Isaac Newton, renowned for recognising the force of gravity in the fall of an apple. Gravity is the force of attraction between all matter. The size of the force depends on the amount of matter. Big things, like the Earth, attract a lot. The Earth's attraction for a smallish apple is a force of one newton (1N). We say that the apple has a weight of one newton (Figure 4.17).

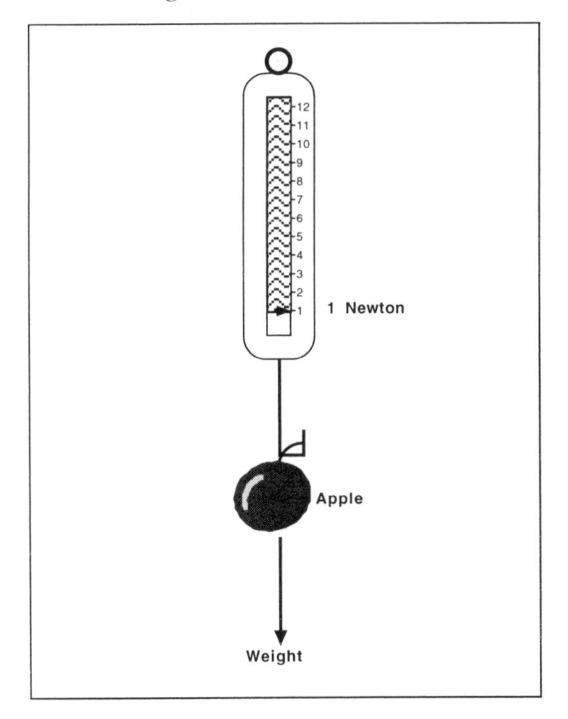

Figure 4.17 Weighing an apple

Usually we talk about the weight of apples in grams, kilograms or pounds. Scientifically these units measure **mass**, the amount of apple 'stuff'. An apple with a mass of 100g has a force of gravity or weight, of 1N. We can see the difference between mass and weight if we take the apple to the Moon. Because of its smaller size, the Moon's gravity is only one-sixth that of the Earth's. The mass of the apple is still 100g, but its weight is only one-sixth of a newton. Pictures of astronauts on the Moon show they are able to move more freely because of their reduced weight.

Forces at work

Forces are usually recognised from their effects. They change the way something moves and they may change something's shape. Road safety provides a good topic to study forces. A car with a large engine can exert large forces and produce a rapid increase in speed, that is a large acceleration. The car may need to have strong brakes to apply the friction to slow the car down again. In a collision, impact forces change the shape of the car body. Cars are now designed so that parts of their bodies crumple in a crash; this reduces the forces on other parts and protects the passengers inside. Further protection of passengers is provided by seat belts. These stretch a little and give a restraining force to stop the passengers continuing to move forward when the car stops suddenly. (See Figure 4.18.)

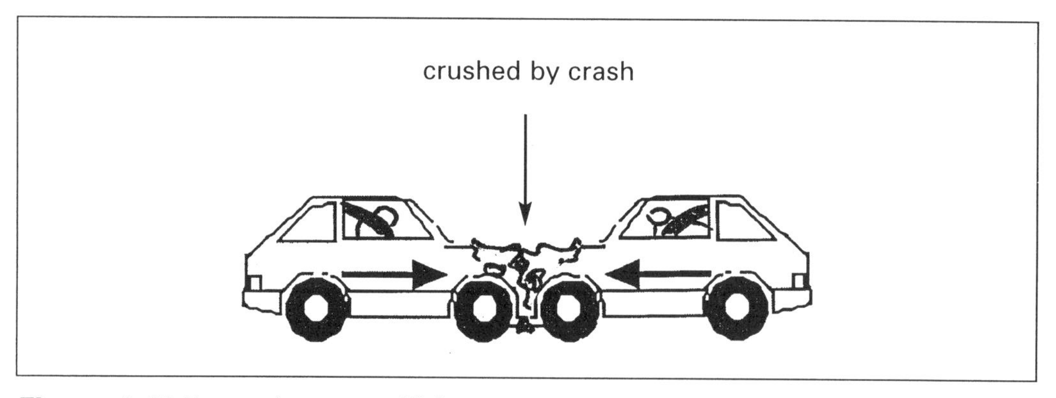

Figure 4.18 Forces in a car collision causing crumpled bonnets and stretched seat belts

The amount that a material changes shape (is deformed) depends on the size of the deforming force – bigger deformations of cars come from harder crashes. Different materials deform in different ways. Seat belts are strong so do not deform much. Also the stretch disappears when the force is removed. This is called **elastic deformation** ('elastic' rubber bands are good at this). By contrast the metal of car bodies will stay deformed even after the force has gone. This is called **plastic deformation** which is confusing as most plastic materials are good at deforming elastically (see Chapter 3, section on 'Mechanical properties').

In order to sort out what kind of movement a force produces, we need a few terms to be defined:

- **speed** is the rate at which something moves its position; it is usually measured in metres per second (m/s) or kilometers per hour (km/h);
- **velocity** is the rate at which something changes its position in a straight line measured in m/s or km/h;
- **acceleration** is the rate at which something changes its velocity, measured in m/s per second, or m/s².

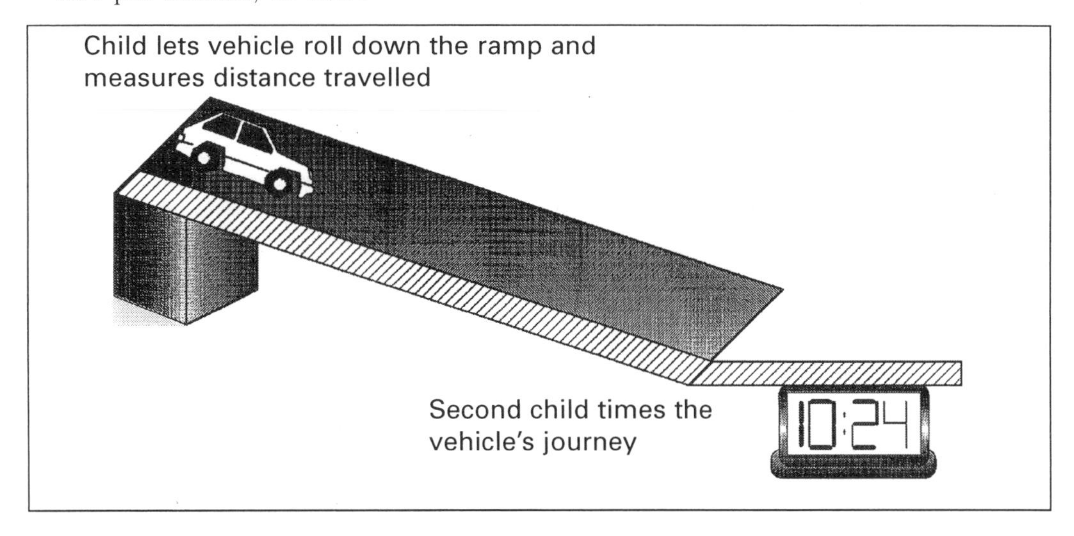

Figure 4.19 Investigating forces causing motion

Investigations with toy vehicles (Figure 4.19) are good ways to see that a force is needed to change the velocity, either by changing its size (faster or slower), or its direction in travelling round corners. It is also clear that greater force is needed when the vehicles have a greater mass. The exact relationship was discovered by Isaac Newton about 300 years ago.

Force produces a change in the direction of the force, which is proportional to the acceleration and to the mass of the object. If the force is in newtons, the mass in kilograms and the acceleration in m/s^2, this can be written as an equation:

force = mass × acceleration.

This is called Newton's Second Law of Motion. It can be investigated using force meters and timers with moving objects.

Newton's First Law of Motion says what happens to the motion of an object when there is no force acting. Either it remains at rest, or it continues to move with steady velocity. The latter state is surprising – everyday experience is that things don't just keep on moving without a push; but that's because there is always some friction or air resistance on Earth – a slowing down force. Think of skiing on very smooth ice – you will not stop until you increase the friction by digging the skates into the ice. Spaceships continue to move steadily without their engines on.

Friction might therefore be considered a problem, slowing everything down – but how would we ever get started without it! Whenever we take a step we are pushing backwards against the ground, so the ground pushes forward on us – even more when you are using starting blocks (Figure 4.20a). This is an example of Newton's Third Law of Motion. This says that whenever an object A pushes on an object B, B pushes on A with an equal and opposite force. In space there is nothing to push off, so how do you get moving? Rockets have to throw their waste out behind so this pushes back (forwards) on the spaceship (Figure 4.20b).

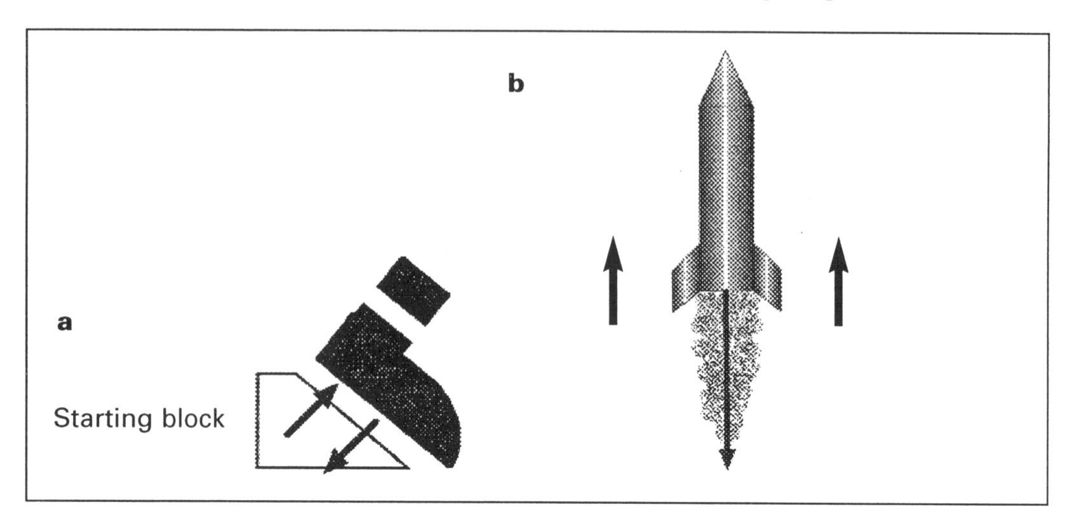

Figure 4.20a, b Changing motion (a) on the ground and (b) in space

Balanced forces

Everything on Earth has the force of gravity acting on it. How then can things remain at rest? They need to be supported, e.g. when books rest on a table, the force of the table upwards on the books just balances the downward force of gravity. If the table is not strong enough to support what is placed on it, then gravity prevails and the books fall to the floor! The science of structures is concerned with both the mechanical properties of materials (see Chapter 3) and forces in balance. If the structure stays up, the forces in it balance gravity!

There is also a balance in the rather different situation of water. The upthrust of the water cannot support the books so they sink until they rest on the bottom. The books can float, however, if they are placed in a boat – what is the difference? The boat has resulted in the upthrust force being large enough to balance the weight of the books (and the boat). Of course the boat sinks lower in the water when it is loaded with books. This it turns out is the secret of floating. The boat displaces – pushes out of the way – more water. You can feel the increase of the upthrust if you try to push a floating object down into water. The ancient Greek scientist Archimedes explained what happens in his principle: the upthrust force is equal to the weight of the water displaced (Figure 4.21).It is an example of Newton's Third Law – when the water is pushed out of the way it pushes back with an equal and opposite force. The amount of water displaced will depend on the volume immersed in the water. Making a boat increases the volume (because it's hollow) so the upthrust is greater (Figure 4.22).

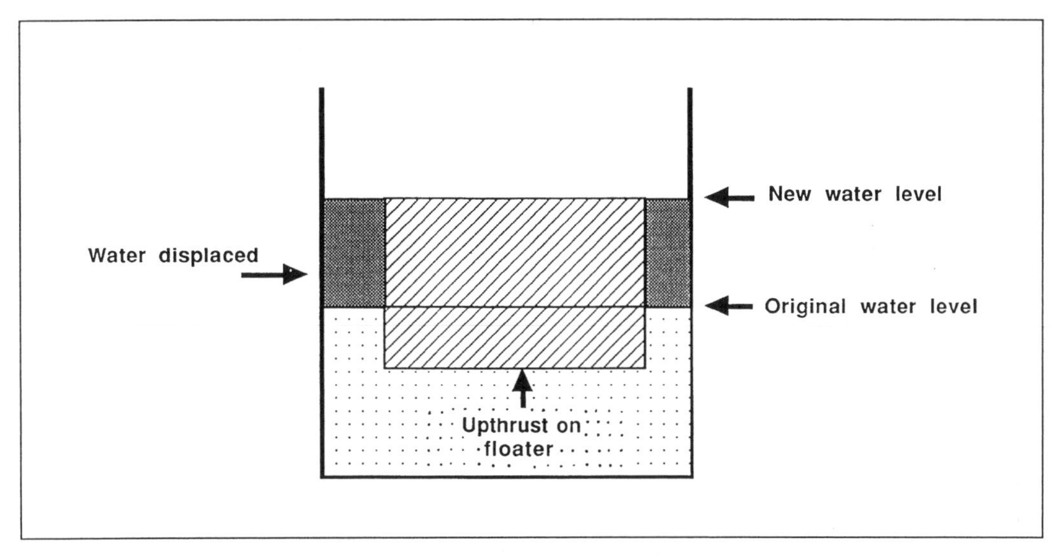

Figure 4.21 Archimedes' principle

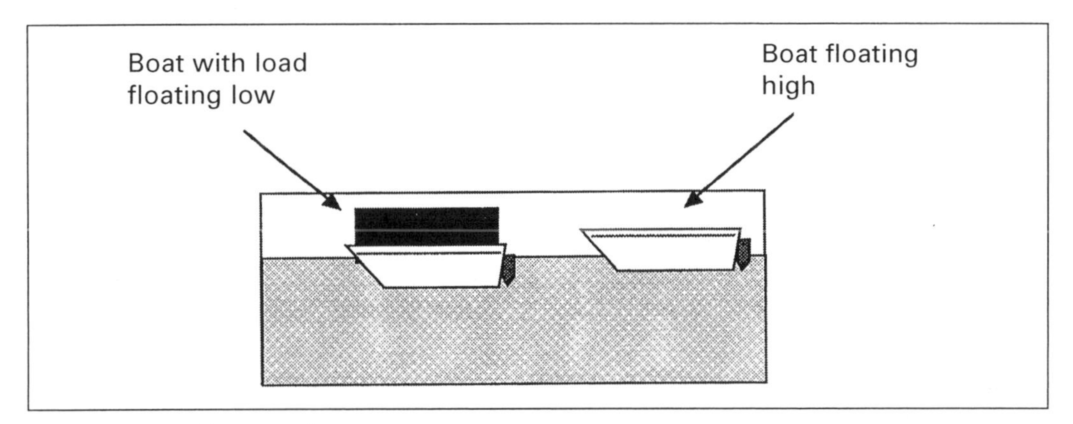

Figure 4.22 Floating boats

Apart from boats and hollow objects, we observe that light things float. To be more specific, things that are light for their size will float. The comparison of mass to volume is called **density**. The density of a material is the mass of a unit volume. For example one cubic centimetre of water has a mass of one gram so the density of water is $1 g/cm^3$. Things which are less dense than this will float because water's upthrust is greater than their weight. Materials which are more dense will sink because water's upthrust is insufficient to support them. Such things will weigh less in water though because there is some support. This explains why very large stones are often moved during floods – they weigh less in water.

Steel has a density seven times that of water, so how can steel boats float? This is because the displacement of the water is by the hollow steel hull of the boat, mostly full of air, so its average density is less than that of water. If the hull is holed and fills with water then the boat becomes more dense than just water, and sinks.

Another way of looking at the situation of floating is in terms of the **pressure** under the water. Pressure is defined as the force per unit area, so it is measured in newtons per square metre or centimetre (N/m^2, N/cm^2). As depth in water increases, so the weight of the water above increases. The pressure at any depth can be worked out as the density of the liquid times the depth. So the deeper something sinks into the water the greater the pressure on its lower surface, and the bigger the upthrust.

What is energy?

Children pick up the idea of energy at an early age, being told their active behaviour shows they have lots of energy. Adults make the connection between what they eat and exercise with comments such as: 'I'll have to play a game of squash to work off this chocolate bar'. This acknowledges several important aspects of energy:

- the chocolate bar is a *source* of energy;
- the human body can *store* energy;
- human activity *transfers* energy;
- the transfer results in the energy being *transformed* into something different – movement and then getting hot.

Similar aspects of energy can be identified in most events, such as:

- climbing a flight of stairs;
- driving a car;
- switching on a torch;
- beating a drum;
- heating a bowl of soup in a microwave oven;
- growing sugar cane in a sunny climate.

The simplest working definition of energy is that it makes things happen, or change, and that this changes the nature of the energy. For example, the operation of the torch changes the chemical energy in the battery first to electrical energy and then to light and heat.

It has been common to identify different types of energy. In the list of events above for example you can find human chemical energy, movement energy, position energy, fuel chemical energy, sound, heat, radiation (microwave, sunlight and heat) and plant chemical energy. These labels are just for convenience and can vary. For example, the energy transformation when you bang a drum appears to be chemical (in body) to sound (in air). However, when the drum is struck it stretches elastically storing position energy, before springing back to vibrate and cause the sound; sound is the movement of the air, so is movement energy. Many types of energy can be labelled as position or **potential energy** and movement or **kinetic energy**. The important concept is that all kinds of energy are interconvertible and all have the same capacity to make things happen.

Money provides a useful analogy for the concept of energy. A person's wealth is made up of many elements including cash, bank deposits and valuable possessions. Each can be converted into the others. Banks and other financial institutions base their business on such transactions, which can be precisely valued in one of the world's currencies – or the ecu! Energy already has an internationally accepted currency. The scientific unit is the joule, commemorating the work of the English scientist James Joule. In pursuit of his aim to unify the concept of energy, he even spent his honeymoon measuring the temperature of waterfalls!

Conservation of energy

It is a basic principle of science that energy cannot be created or destroyed but only transferred. Accounting for energy in these transfers is like financial accounting. When the £50 withdrawn from the cash dispenser has gone it is possible to identify where it went – even if we find it hard to believe that it has all been spent! The conservation of energy means that we can count the joules which are input into a change and expect to have the same amount after.

Why then are we advised to 'save energy' and warned that 'the world may run out of it'? What the world may run out of is certain sources of energy – the fossil fuels gas, oil and coal – as these take millions of years to form. They provide very convenient energy sources because burning them releases a large amount of energy which can be transformed into electrical energy in power-stations. Of course it is expensive to use such fuels, but what happens to the energy they release?

Think about cooking a meal in an enclosed kitchen. The energy from the cooker warms the room as well as cooking the food. Here it is less useful, and will eventually spread throughout the house and out through the windows. In this form it cannot be controlled in the way you can control the energy from a cooker. It is useful, controllable energy which is lost, not energy itself. The effects are potentially serious because most of the energy is obtained from burning fossil fuels. This releases carbon dioxide which collects in the atmosphere to form a barrier to the spread of this uncontrollable energy. Energy can still pass into the atmosphere from the Sun, but it is being prevented from leaving, like in a greenhouse. The result is that the Earth is expected to slowly warm up, changing the climate and melting the polar icecaps.

Measuring energy

For health reasons many people monitor their food intake, including energy. Table 4.1 is from the food packaging of chocolate biscuits and shows energy values in kilojoules (kJ) and kilocalories (kcal), i.e. in thousands of joules and calories. Dieters have traditionally 'counted the calories', which is the older British unit of energy. Counting joules may sound more attractive but there are more of them – a simple calculation from this packet shows 1 cal is approximately 4.2J.

Fats and carbohydrates are the foods which we eat for their energy content. Foods get their energy from the photosynthesis of plants (see Chapter 2), e.g. a 100g of banana provides about 300kJ, 100g of sugar provides 1,700kJ and 100g of alcohol

2,900 kJ (which is why alcohol is used in Brazil to fuel cars instead of petrol!).

Joules and calories are both rather small units, and humans use millions of them everyday. Table 4.2 shows the values in megajoules (MJ) of some of these uses.

Table 4.1 Nutritional information from a packet of chocolate biscuits

Average values	*Per 100 g*	*Per biscuit*
Energy	212 kJ	367 kJ
	506 kcal	88 kcal
Protein	6.8 g	1.2 g
Carbohydrate	65.7 g	11.4 g
of which sugars	29.1 g	5.0 g
Fat	24.0 g	4.2 g
of which saturates	12.1 g	2.1 g
Fibre	2.3 g	0.4 g
Sodium	0.5 g	0.1 g

Table 4.2 Energy values

Energy content of:	*Energy in MJ*
Output per second of a large power-station	2,000
Stored in a human body	300
In 4.5 litres/1 gallon of petrol	170
Used by an average British family in 1 day	150
Daily food requirement of 15-year-old boy	12
Daily food requirement of 15-year-old girl	10
Energy for a hot bath	4
A day's heavy manual work	4
Mars bar of 100 g	2
Sleeping for 8 hours	2
Walking 5 km	1

Heat and temperature

These two words are often used in everyday language to refer to aspects of the same thing. In science the relationship is carefully defined. Heat is the cause of a change; temperature is the effect. Temperature measures the 'hotness' of an object which contains energy or heat.

What is called heat, is really two kinds of energy. There is energy in transit (such as from the Sun), and energy in hot things (such as a bowl of hot soup) or **internal energy**. Energy in transit is usually **radiation energy**, or part of the electromagnetic spectrum (Table 4.3). This consists of different types of energy waves all of which travel with the speed of light. The way in which their energy affects the matter on which they fall depends on the type of wave. The most energetic waves are the photons of gamma and X-rays, which can damage the inside of bodies. Next is ultraviolet radiation, which will darken the skin and can produce skin cancer, followed by visible light, the only part of the spectrum that our eyes are designed to detect. Some animals can detect infrared radiation, which we usually call heat rays. Next are microwaves and finally radio waves, which have no effect on our bodies.

Table 4.3 Electromagnetic spectrum

Type of wave	*How detected*
gamma rays	photography, ionising, e.g. radiotherapy
X-rays	photography, e.g. in medical diagnosis, ionising
ultraviolet	photography, fluorescent, e.g. in disco lighting
visible light	the eye, photography
infrared	heating effect, e.g. on the skin
microwaves	electrical circuits, e.g. in microwave cooker
radio	electrical circuits as part of a radio receiver

When radiation is absorbed by matter it is called **internal energy**. It causes the particles of which the matter is made to move faster. This can result in a number of effects. Any object will get hotter and will probably expand. Eventually, if it is a solid like ice it will melt, and the water will evaporate to steam. Other solids may burn or change permanently in other ways (see Chapter 3).

Temperature measures the hotness of an object. When energy is transferred to an object its temperature rises and when it loses energy its temperature falls. When two objects are in thermal contact with each other (not insulated) energy will always flow from the higher temperature to the lower, until the temperatures are equal.

Different types of material increase their temperature at different rates when they absorb energy. For example, water can absorb a lot of energy without changing its temperature very much; it is said to have a high thermal capacity. This fact has important implications for Earth's climate – seas have a moderating influence on temperature change. Iron heats up more quickly. Small objects will heat up faster than large ones. Sparks from sparklers are small iron fragments with temperatures up to 1000°C, but they contain less energy than a cup of boiling water – and will therefore do less damage it they fall on the hand.

Temperature is a measure of how the particles move in an object, not of how much heat or internal energy it has.

Power

Power is the rate at which energy is transferred. A powerful car can accelerate faster than a car with less power because its engine transfers the energy from the fuel to movement at a faster rate. Power is measured in watts (W); a watt is the energy transfer of one joule per second so:

power [W] = energy [J]/time [s].

Light bulbs usually have powers of up to 100 W, whereas heaters are usually rated at 1, 2 or 3 kW (kilowatts). Heating costs more than lighting because energy is transferred at more than ten times the rate. Electrical energy is charged by the unit, which is the kilowatt-hour: using a 1 kW heater for an hour uses one unit of electricity. This equals:

$$1000\,\text{W} \times 60 \times 60\ \text{s} = 3{,}600{,}000\,\text{J or } 3.6\,\text{MJ}$$

Depending on the tariff, this costs about 6–7 p at present. It is about the same amount of energy that a person would transfer if they did a day's physical work (see Table 4.2). This shows just how we take for granted our modern energy-rich society, and why we are using so much fossil fuel in power-stations.

Forces and energy

Energy is defined as the capacity to do work. This sounds like everyday language, but in science the word work has a specific meaning. It is what is done by moving against a force, e.g. if you lift an apple, weight 1 N by 1 m, you have done 1 J of work (Figure 4.23):

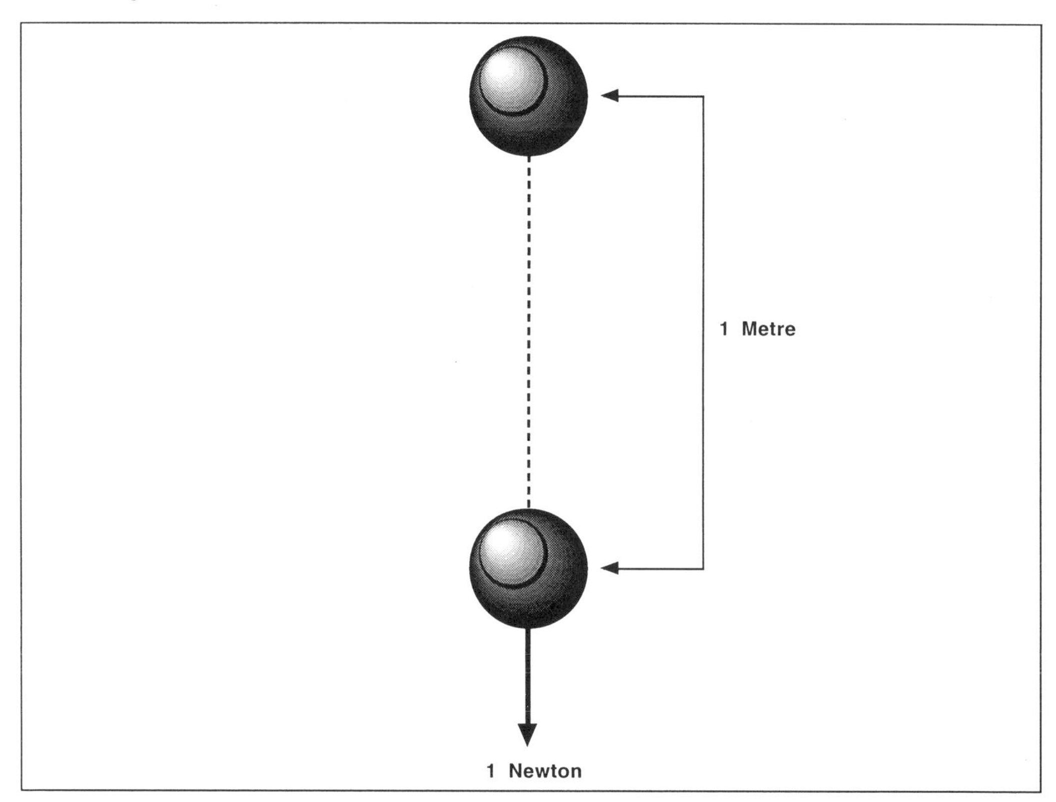

Figure 4.23 Doing a joule of work

work [J] = force [N] × distance [m]

Lifting a 5 kg sack of potatoes on which the force of gravity is 50 N by 2 m, to put it on a shelf requires

50 N × 2 m = 100 J of energy

The potatoes now have 100 J of potential energy – which they could release by falling on you!

Force can be thought of as an agent, making movement or shape changes. Energy by contrast is about the condition of things. When the condition changes then an energy transfer can be accounted for.

Children and forces and energy

Knowledge and understanding

Research into children's ideas of energy has been quite extensive. It shows that young children associate energy with life and with activity – and themselves! It is a thing which is gained (e.g. from food) and then lost. This mirrors the development of the concept in science. It was not until the work of Joule in the nineteenth century that energy was understood to have a more general application leading to the idea of conservation of energy.

For children, force is also associated with their own actions or those of living things and not with inanimate objects except those like cars which can move themselves. Force is seen as strongly connected with movement, and to 'run out'

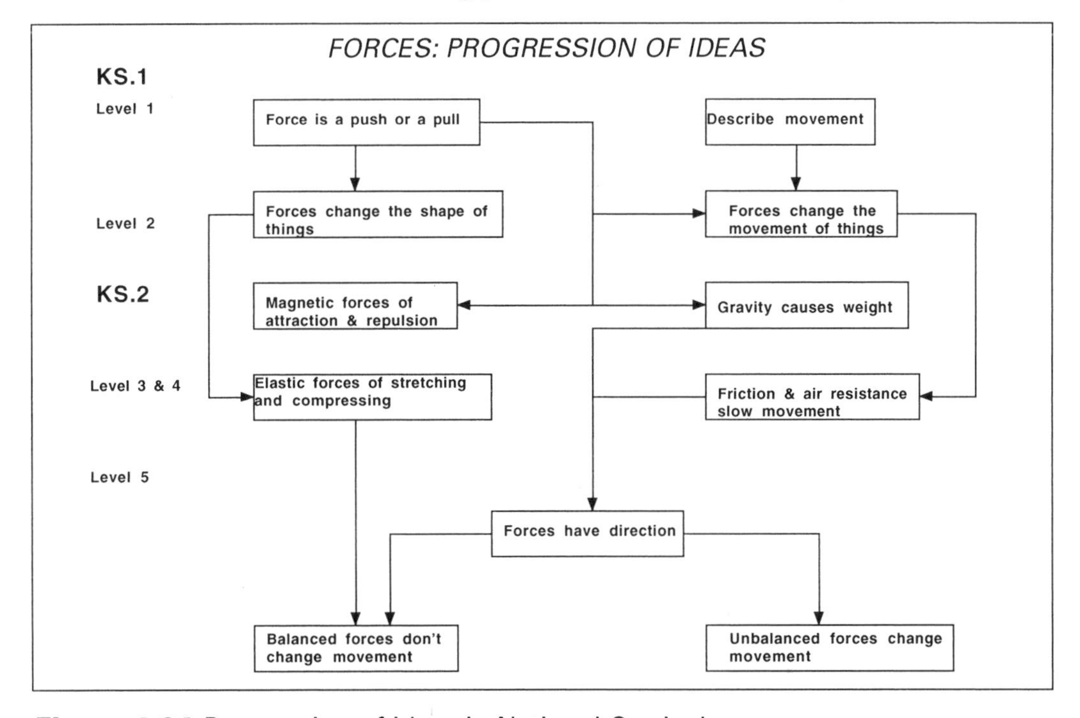

Figure 4.24 Progression of ideas in National Curriculum

when movement stops. The notion of a force as a push, pull or combination is readily accepted when connected with movement, for example in a fight. However, a balance of equal forces, as when a person stands still, is less easily understood.

In many situations, as we have seen in the previous section, force and energy descriptions are complementary. For this reason the National Curriculum for England and Wales (though not for Scotland and Northern Ireland) avoids energy – but children and the books they read won't!

Much of the challenge, as pupils progress through primary school, is to help them differentiate between movement, force and energy (Figure 4.24). Experiences need to be planned which develop their use of the words in a more precise, scientific way. Their ideas can be challenged by situations and events which are counter-intuitive, such as heavy objects falling as fast as lighter ones, and of wrapped cold things staying colder than unwrapped ones (see Figure 4.25).

Figure 4.25 Concept cartoon of wrapping a cold object

The goal for the end of the primary stage is to enable pupils to begin to look at the world through force or energy 'spectacles', as a preliminary to the secondary school treatments of Newton's mechanics and the kinetic theory of matter.

Skills and attitudes

There is a need to ground the abstract concepts with lots of hands-on experience, and opportunity for communicating with the language of observation – fast and slow, pull and push, hot and cold, etc.

There are good opportunities for predicting and testing hypotheses, e.g. about floating or the bouncing of balls. The results may not be conclusive but that's the lesson of science: 'The destroying of a beautiful theory with an ugly fact'!

The practical work can often involve measuring – with scope for a range of recording methods, including tables and graphs.

Teaching forces and energy

Questioning

This is a key skill to develop with pupils, to elicit their ideas at the beginning of a piece of work. Most of the key ideas, e.g. floating and sinking, will be revisited several times during primary years, and so it is important to ensure continuity as well as progression for the pupils. The following case study shows a teacher using questions with a group of 5 and 6 year olds. The prompt question is 'How can we make the lemonade bottle sink and the stone float, in this tank of water?' The teacher uses questions to probe their understanding and to encourage the children's hands-on exploration.

The activity was set up in an open plan classroom with the teacher and six pupils at the water tank. On a table there were items the teacher thought the pupils might want to use – elastic bands, shells, strings, plastic cups, Plasticine. The rest of the children were directed to other activities.

Teacher What is this?
Child 1 A stone.
Teacher How does it feel? Pass it round.
Child 1 Hot.
Child 2 Cold.
Child 3 Warm.
Teacher What is this?
Child 4 A bottle.
Teacher Pass it round, feel it.

Bottle falls into water.

Teacher What is it doing?
Child 1 Floating.
Teacher What will the stone do in water?
Child 6 Drown.
Teacher Will it drown?
Child 5 Sink.
Teacher We call it sinking, not drowning, when it goes to the bottom of the water. So what do you think will happen?
Child 2 Sinking.
Teacher I've got a special job for you. I want you to try to make the bottle sink to the bottom of the water tray like the stone does. How can you do that?
Child 1 Take the top off the bottle!
Teacher What will happen then?
Child 6 It might sink.
Teacher What do the rest of you think?

Children Sink, float, stay on top.

Children push neck of bottle into the water and a small amount of water goes in. The children put the top back on.

Teacher What is happening?

No response as bottle is partially submerged.

Teacher Is it floating or sinking? Look through the side of the water tray. Is it on the bottom?
Children Floating, sinking.
Teacher Has it gone right down to the bottom?
Child 5 No.
Teacher What can we do?
Child 1 Put more water in.

The children push the whole bottle under the water with the top off to fill it up.

Teacher What is coming out?
Child 4 Bubbles!
Teacher Where have they come from?
Child 3 The bottle!

Children put the top on again to see if the bottle will sink. It is still only partially submerged.

Teacher What is in the bottle?
Child 2 Water.
Teacher What about there?

Points to the large pocket of air.
No response.

Teacher Why is that part of the bottle up and that part down?
Child 5 Because there is no water in there.
Teacher What happens if you push the bottle down to the bottom?

Children push the bottle down and let it go.

Child 1 It comes back up.
Teacher I want the bottle sitting on the bottom. How can you make it stay down?
Child 5 Pour more water in.
Child 6 Take all the water out.
Teacher If you take all the water out the bottle will float again won't it?
Child 6 Oh yes.

Meanwhile, other children have fetched a cup and funnel to help them fill the bottle.

Teacher Is that full now?
Children Yes.
Teacher What's going to happen now? Let's think about what will happen.
Children It will sink!

The bottle sinks to the bottom.

Teacher It does – you clever lot!
Teacher Now I want you to make the stone float.
Child 4 Put it in the cup.

The one they used to fill the bottle.

Child 1 No, it is too big, it will tip up.

Stone tips cup and sinks.

Child 3 I know, we need something bigger.

Children go and fetch a larger container and put the stone in.

Children It's floating!
Teacher Well done, it's floating.

Safety

The hazards of work on this topic are obvious and familiar – heavy, fast-moving or hot things. Beware, the familiar is often overlooked, after all, most accidents happen in the home.

Using appropriate contexts

The world is full of forces, movement and energy; the skill is to choose those which give the right combination of familiarity and challenge, a combination which shifts to the latter as children get older.

Forces in the nursery
Every time a child kicks a football or rides a bike he or she uses a force to make movement. For very young children connecting the foot with the ball is quite an achievement, but with practice the force can be controlled. Adults can celebrate the increase in proficiency by talking about how far the ball travelled with a hard kick. The language used can support the child's understanding.

Splash painting is a fun way to use forces – the harder you flick the further the paint goes (but remember to wear shower caps to protect the hair!). A suitable choice of colours can also help children learn about colour mixing. An alternative is to use a lever and suction pump in a bowl of water to produce a big spray with a strong push. Plant spray bottles can be filled with water and food colouring and patterns and colour mixes sprayed on an old sheet. Or provide a bowl of water with wet sponges to aim at a target . . . on an outside wall.

Another activity for outside is to use a pulley to raise a bucket of water, or sand. The pulley effectively magnifies the force applied. Without the pulley the bucket is too heavy to lift. Let the children tackle the problem. They can raise the bucket by emptying out some of the contents to make it lighter, or they can work out how to use the pulley. This kind of knowledge can then be applied to similar situations.

Gravity can also be used as a force to have fun with. A ramp made from plastic guttering becomes a runway for cars or balls to whiz down. Funnels can be attached to tubes and poked through the holes of a milk crate, so that children can pour sand or water through and watch it fall into the sand or water tray.

There are many similar activities which use everyday objects to explore forces and motion. Knowing something of the scientific principles can help the teacher to be creative in the design of activities, to further the children's understanding. It can also give the teacher the confidence to share the ideas with parents so that they see the educational value in what might otherwise appear to be a random, messy play session.

Forces for older children

Older pupils are stimulated by more exotic contexts. Space travel and the problems caused by the excessive use of fossil fuels, can appeal in different ways. Another way to provide an engaging context is through a story. For example, in *Spacebaby* by Henrietta Brandford, an alien visits Earth to solve the problem of a loss of gravity caused by it leaking. There are several books which cover floating and boats in an attractive way at a range of levels (see resources list in Appendix).

Developing understanding

Misconceptions can be challenged by taking children through conversations about inconsistent concepts, for example:

- 'Energy is about keeping fit. It makes you run fast' (8-year-old) – if energy is connected with both movement and life and exercise is good for healthy life – does exercise give you energy or use your energy up?
- 'The ball stopped when it ran out of force' (10-year-old) – why does a kicked ball behave differently when it travels through the air from when it falls on a muddy field – talk about the forces.

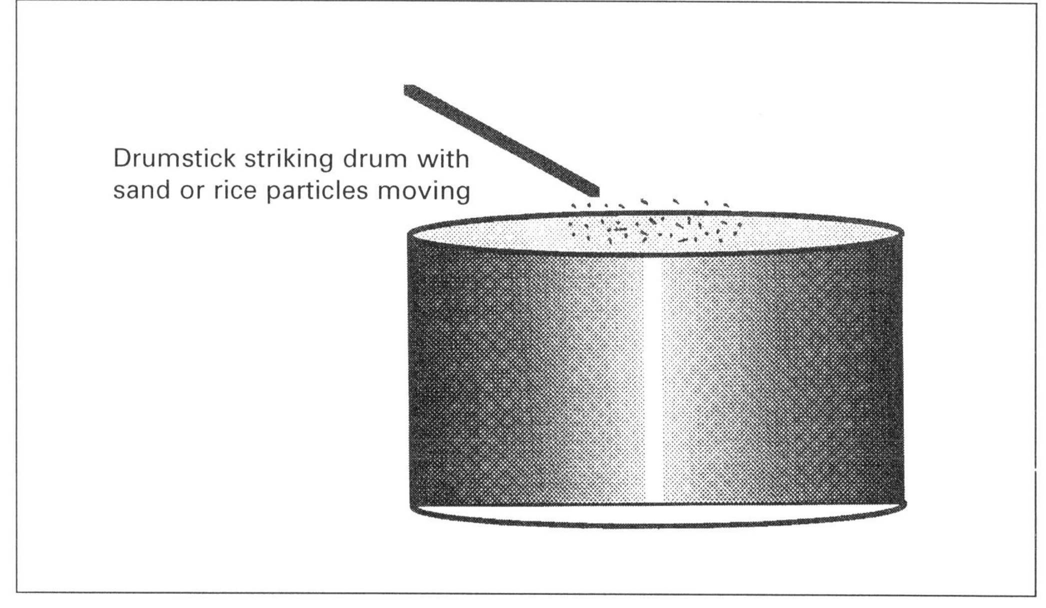

Figure 4.26 Vibration of a drum

SOUND AND LIGHT

Understanding sound and light

Sound and light are attractive and accessible topics for children which can involve both careful observation (using the senses) and the use of sound and light in music and art. Both sound and light can be thought of as energy transmitted by wave motion, but this concept is not required for pupils until the secondary stage.

Sources of sound

All sound has the same origin – a rapid to-and-fro movement called a vibration. It can be seen by scattering rice or sand on the surface of a drum skin or the sound box of a guitar (Figure 4.26).

Musical instruments are grouped by the way they make their sounds:

- **percussion** – striking a solid surface;
- **string** – plucking, strumming or bowing a tightly stretched string or wire;
- **wind** – blowing down a tube.

The voice operates like a wind instrument – air in the windpipe makes the structures of the larynx in the throat vibrate.

A force is needed to make a sound. Increasing the force increases the size or amplitude of the vibration. This means the energy or level of the sound wave is greater and it is heard to be louder. As sound travels from a source it spreads out so the energy at any point falls with distance, which is why the sound gets softer. Musical instruments are designed to vibrate at different rates, or frequencies. This affects the pitch of a note. For example the note middle C has a frequency of 256 vibrations per second. Going up an octave doubles the frequency.

Travelling sound

'In space no one can hear you scream' said the trailer of a sci-fi movie, and it's true. Sound needs something to vibrate (a medium) for it to travel. Its failure to pass through a vacuum can be demonstrated by putting a ringing alarm clock in a jar and pumping out all the air. As the air leaves the bell ring fades away. Sound travels best in solids – which is why musical instruments have sounding boxes. It also travels well in water – whales and dolphins have extensive communications there. In air it is not so effective but works the same way. The vibration pushes the particles of air one way then another, so they are compressed and 'decompressed' or rarefied. This motion passes through the air and when detected by the ear, we hear. A model of a travelling sound wave is that of a 'slinky' spring. The to-and-fro movements of the coils are similar to those of the air particles (Figure 4.27).

The wave is usually represented by plotting the pressure against the distance, at a particular moment. This gives a curve, looking like a ripple on a pond, in which the features of the sound are represented. Figure 4.28 shows the curves for two sounds, that in **(B)** is louder and of lower pitch than that of **(A)**.

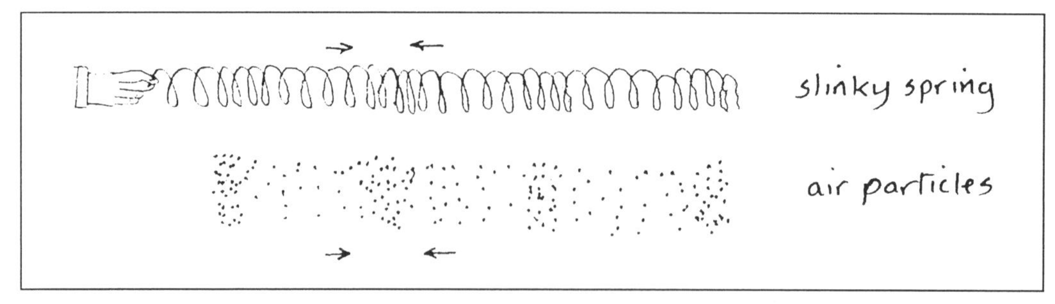

Figure 4.27 Model of sound waves

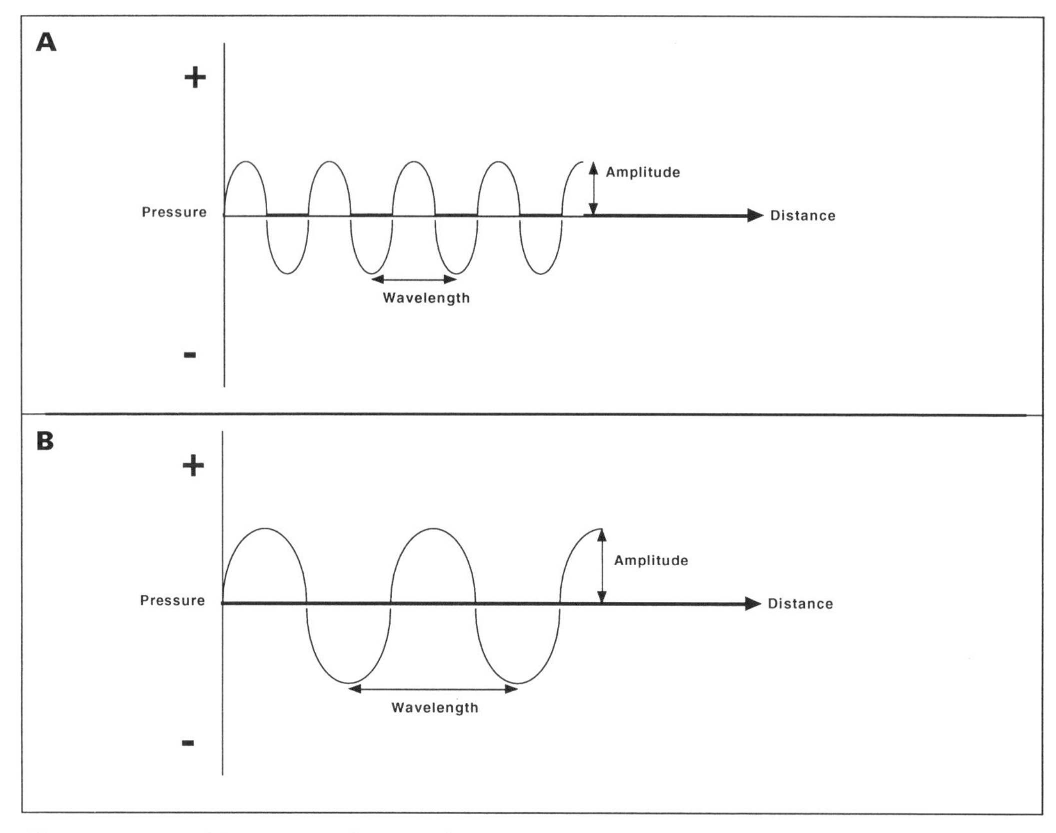

Figure 4.28 (A) and (B) Graph of sound wave

The qualities of a sound depend on these features, as follows:

- **loudness** – depends on the height of the ripple, or amplitude;
- **pitch** – depends on the frequency, but also the wavelength;
- **speed** – depends on the medium; it is given by:
  speed of wave [m/s] = wavelength [m] × frequency [Hz].

When sound meets a surface it can be reflected and/or transmitted through the different medium. Ultrasound has a frequency too high for humans to hear, though bats use its reflections to navigate in the dark. Humans use it to picture the inside of the body, e.g. for checking the development of a foetus. If the reflected sound comes back to a listener a second or more after the original sound, an echo is heard. The science of acoustics is used in designing concert halls so that there is the right amount of reflected sound and no echo. Noise reduction involves the use of sound absorbent materials, such as foam pads in ear muffs or air in double-glazed windows.

Hearing

The ear is a remarkable device, which has:

- accuracy – a false note can be detected if its frequency is slightly different;
- selectivity – listening to a quiet conversation at a noisy party;
- speed – listening to what you say as you say it;
- range – from a frequency of about 50 to 20,000 Hz in young people;
- sensitivity – the loudest sound you can hear without damage is a trillion (10^{12}) times louder than the softest.

The ear can be damaged by exposing it to even louder sounds, a risk run by many young people at discos and pop concerts. Sound levels are measured in decibels (dB). This is a logarithmic scale, which means that if the sound level increases by 3 dB, the energy is doubled. Table 4.4 shows the loudness of some familiar sounds and the Health and Safety Executive recommended maximum exposure times to avoid ear damage. The risks are clear!

Table 4.4 Sound levels

Sound	Loudness at 1m from source (dB)
Ticking watch	20
Conversation	55
Background music	65
Alarm clock	80
Wembley pop concert	Up to 105 at 50m
Disco sound system in club	Up to 120

Loudness (dB)	Maximum safe exposure time (h)
90 (conversation difficult, need to shout)	8
93	4
96	2
99 (conversation possible only at top of voice)	1
102	0.5
105 (conversation impossible)	0.25

These are the stages in hearing:

- outer ear collects the sound (ear trumpets can collect more sound if needed);
- sound vibrates ear drum, a delicate piece of skin that easily ruptures;
- ossicles act like bone levers to amplify the vibration;
- oval window transmits vibration to cochlea;
- liquid in cochlea moves past hair-like nerve cells;
- electrical impulses from these cells pass down auditory nerve to brain.

This is the simple version – the full story of how the ear works is still being researched! See Figure 4.29 for a diagram of the structure of the ear.

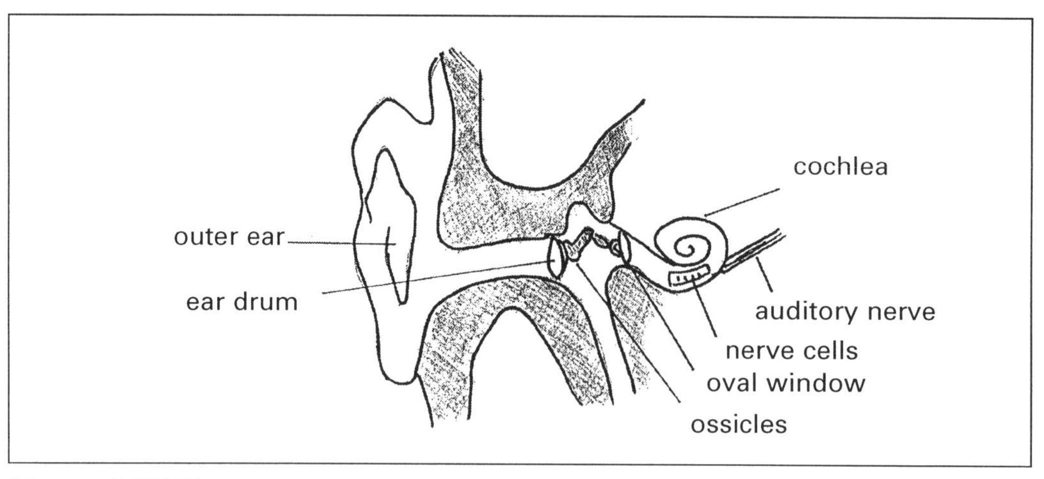

Figure 4.29 The ear

What is light?

This is not an easy question to answer. In many ways it is a wave, like sound, which radiates energy from a source, but it is quite different from sound because it can travel through a vacuum. So light is not waves of stuff but just energy in transit. Light is that part of the electromagnetic spectrum (Table 4.3) that we can see. Electromagnetic waves are emitted by hot objects; how hot the objects are determines what waves are emitted, e.g. the Sun emits infrared and ultraviolet radiation as well as visible light. Infrared radiation produces heating and ultraviolet radiation can tan skin or even cause cancer of the skin. Hotter stars send out X-rays and cooler ones emit microwaves and radio waves.

There are ways in which light behaves like a stream of particles or packets of energy. The size of the packet of energy depends on the frequency of the light, e.g. ultraviolet affects the skin because its packets are more energetic than visible light; you won't get a tan (or skin cancer) sitting under glass which absorbs ultraviolet.

Travelling light

It is our experience that light travels instantaneously. Careful measurement has shown that it actually travels very fast – about 300,000km/s. Einstein's famous theory of relativity, which revolutionised physics at the start of this century, is based on the idea that nothing can travel faster than light. So the nature of light is quite a puzzle!

Light radiates out from a source in straight lines, but although we use light to see things, we can't actually see light. Light beams and rays can only be seen when the light reflects off something in the beam, such as particles of dust or smoke – hence the dusty disco scenes!

When light falls on an object it can be absorbed, reflected or transmitted, or some combination of these.

Opaque objects absorb light, transparent ones transmit it and translucent ones, such as frosted glass, transmit some. Because light travels in straight lines, opaque objects should cast shadows with sharp edges – the light can either get past it or it can't. However, we often observe soft-edged shadows because light is coming from many different places. For example, on a cloudy day light from the Sun is reflected or scattered from clouds.

A mirror reflects light in a regular way because it has a very smooth surface. It is usual to represent rays of light as straight lines. These reflect off a flat surface at equal angles to it. The law of reflection states:

the angle of reflection = the angle of incidence

where the angles are measured as shown in Figure 4.30. This means that the law can be used for surfaces which are not flat.

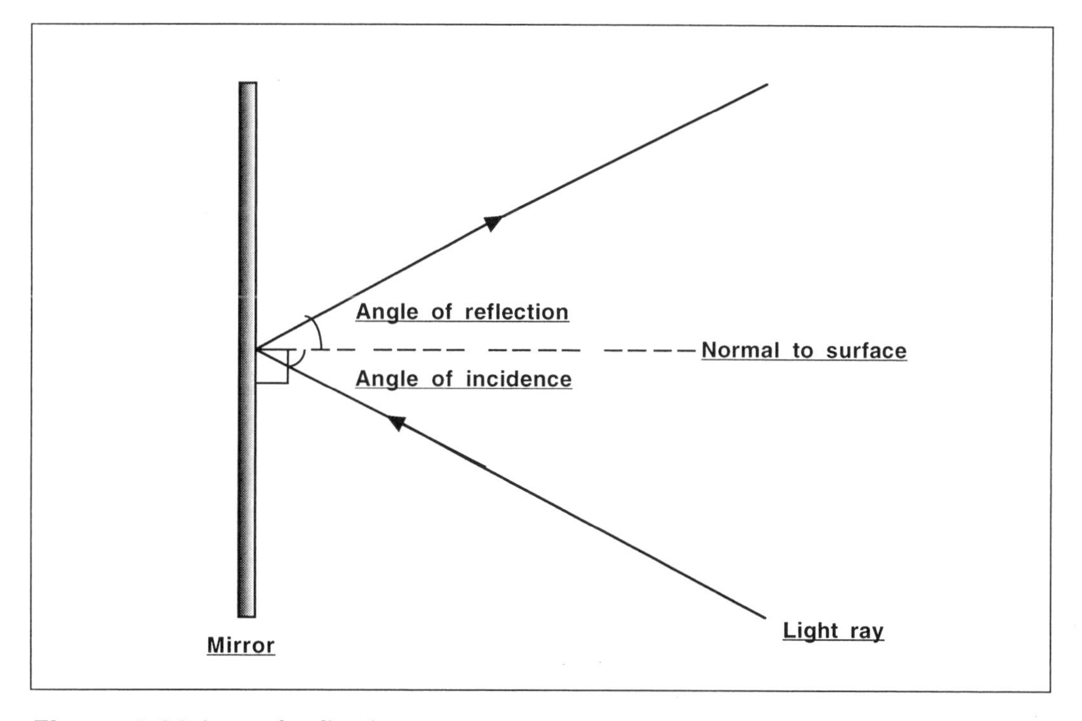

Figure 4.30 Law of reflection

Because of the regular way in which mirrors reflect light they create images. This means we can see things in places where they are not. In a flat (plane) mirror the image is the same size as the object and as far behind the mirror as the object is in front of it. This can be proved by considering the geometry of the reflection of rays from the object (Figure 4.31). In curved mirrors the images are distorted because of the curvature of the reflecting surface.

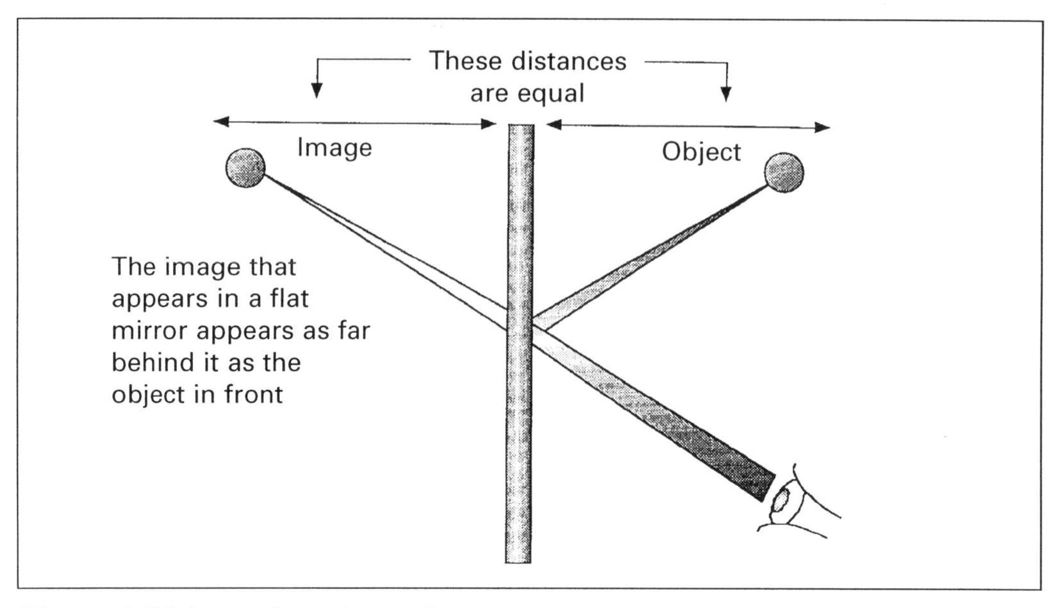

Figure 4.31 Image in a plane mirror

When light is transmitted through an object or medium it changes its speed. If the light travels through at an angle then the direction of the light is changed – the light is refracted. Because of the way our brain interprets what we see, this gives rise to images, such as the apparently broken pencil in a glass of water, and the apparent shallowness of water (Figure 4.32).

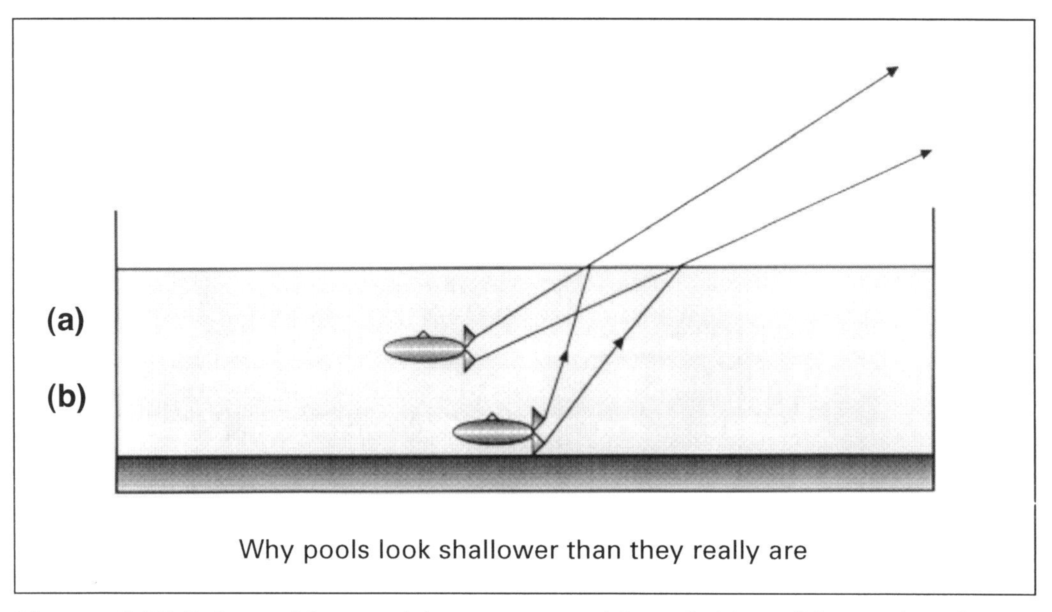

Figure 4.32 Refracted image: (a) apparent position of object; (b) actual position of object

Curved shapes of transparent medium, such as glass and perspex, called lenses can produce magnified or diminished images, and these are used in optical instruments such as cameras, microscopes and telescopes.

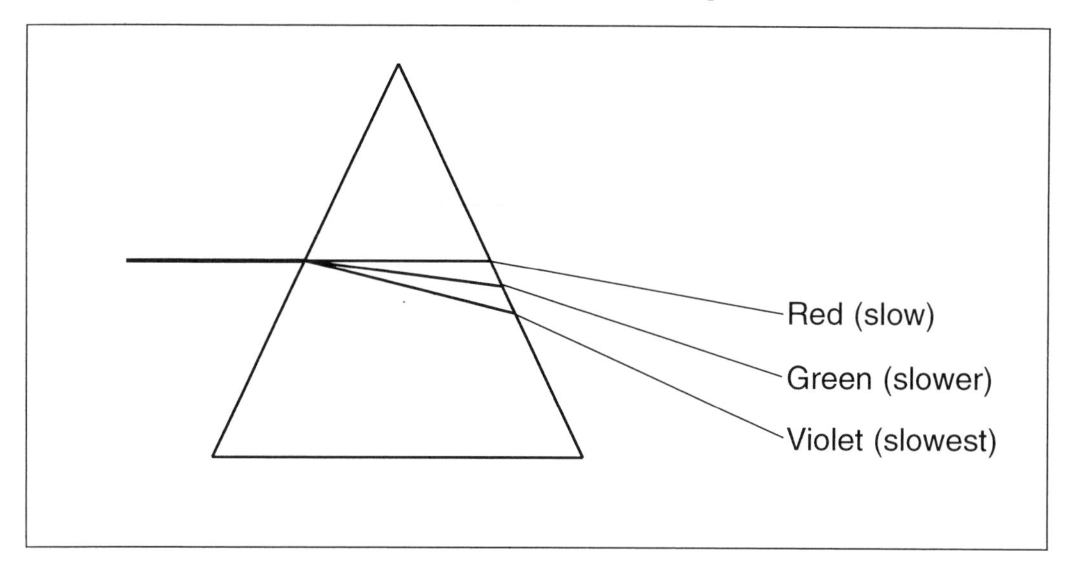

Figure 4.33 A prism makes a spectrum

White light consists of a spectrum of different coloured light. This was discovered by Isaac Newton by using a prism. When the light is slowed down entering the glass, the different colours are each refracted in different directions and a spectrum of colour is produced (Figure 4.33). A rainbow is the result of raindrops acting as prisms to the Sun's light.

Newton proved that the colours came from the white light by recombining them with a second prism. You can make white light from just three colours in the spectrum: red, green and blue (Figure 4.34a). These are called the primary colours, and dots of these can be seen if you look carefully at a TV screen with a magnifying glass. Most of the colour that we see, however, is reflected from objects, and mixing gives different results. A red tomato looks red because when white light falls on it only red is reflected; the other colours are absorbed. If green or blue light illuminates it, the tomato will look black because there is no red light to reflect. The tomato contains a red pigment, and to paint a picture of a tomato we need red paint which also contains a red pigment. We might choose to paint a green tomato with a green pigment; when illuminated this will reflect green light. If, however, we mix the green and red pigments we will get a black tomato because the combination will absorb all colours of light. This is called colour combination by subtraction (Figure 4.34b), because each pigment takes away some of the colours from white light. The same effect occurs with colour filters – a combination of green and red will stop all the light. Notice the difference from mixing coloured lights, which is adding colours – here you eventually get white light, as well as the surprising red + green = yellow.

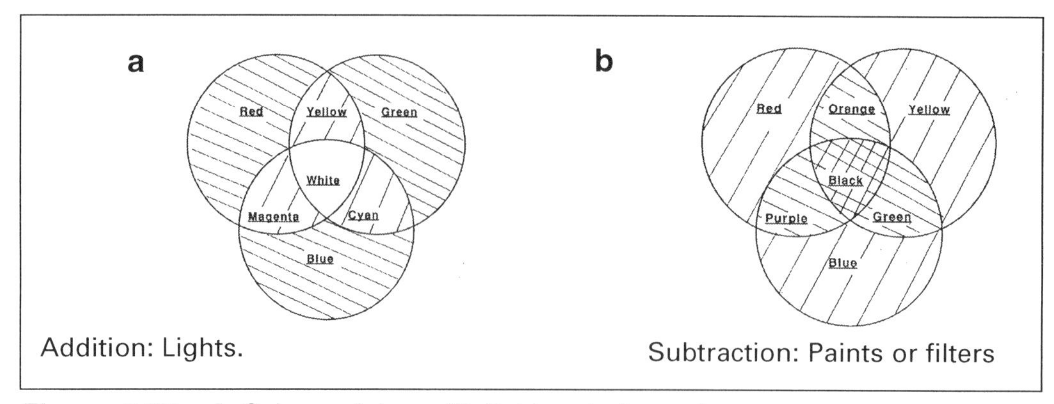

Figure 4.34a, b Colour mixing with light and pigments

Seeing

For most humans sight is the sense with which we first interpret our surroundings – there is a large part of the brain which helps to interpret the nerve signals from the eyes. Light enters the eye through the cornea, the transparent front part of the eyeball (Figure 4.35). This is protected by eyelids and can be lubricated by tears. Some focusing occurs here and the rest is done by the lens behind it, which can be made more powerful to focus on close objects and less powerful for far objects. The ciliary muscles squeeze or stretch the lens – you can feel this if you quickly change your view from far to near. There is a variable aperture to the lens called the pupil, surrounded by the iris, the coloured part of the eye, which controls the amount of light entering. You can see the iris grow, and the pupil shrink, if you shine a bright light into the eye. The light passes through the transparent fluid which fills the eye and an image is formed on the light-sensitive retina. This consists of two sets of receptor nerves, each of which sends information to the brain down the optic nerve. At normal illumination levels the 'cone' cells detect colour; when illumination is low only the 'rod' cells work, and we see in monochrome or black and white.

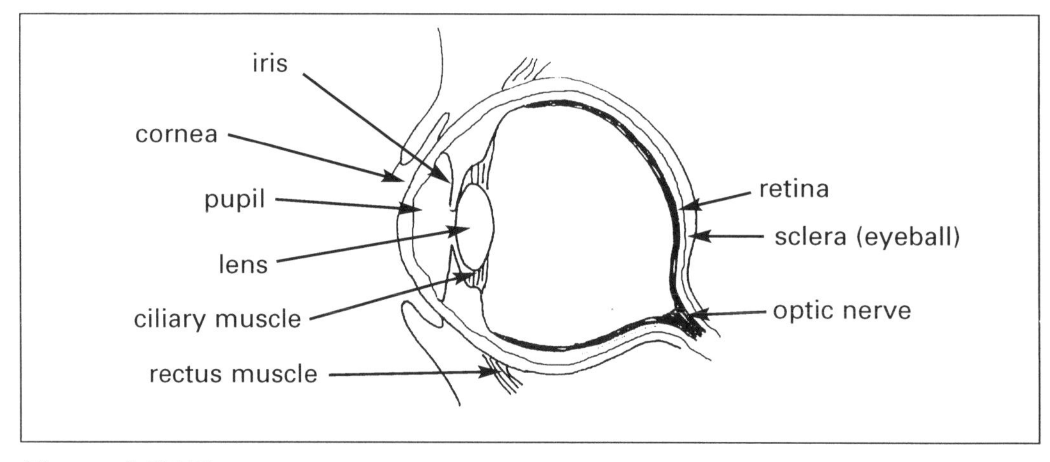

Figure 4.35 The eye

Some people have impaired colour vision; red and green confusion is quite common in males. More common are impairments in focusing: short-sightedness (lens too strong to focus far objects) and long-sightedness (lens too weak to focus near objects). Both can be corrected by spectacle lenses. In middle age the lens loses its flexibility and people's vision is impaired for both far and near objects, so bifocal lenses can be worn.

Children and sound

Knowledge and understanding

Children will have experience of making sounds (even music perhaps!), from an early age. Younger children will associate the sound with the object and their own actions in making the sound. They need to experience a wide range of sound producers, so they can see patterns in the kinds of sounds. They can be introduced to the concept of a vibration by sprinkling sand onto a drum or sound box or by dipping a ringing tuning fork into water so that ripples are generated. A plastic or wooden ruler can be flicked while held on the table edge to show variations of loudness (size of vibration), and pitch (length of ruler free to move related to frequency of vibration).

Children do not readily think about sound travelling, and when they do they can have difficulties with the experience of sound travelling round corners, through doors, etc. (in contrast to light). Listening to a string telephone or to the end of a string from which a fork is suspended can focus attention on this aspect. The fact that sound travels better in solids and liquids than in air surprises children.

Many children have an 'active ear' model of hearing where the important thing is the hearer's attention rather than the sound source. The concept of the vibration as the source of sound is probably essential for progress in these areas.

The goal is for pupils to provide explanations of the behaviour of sound which will be consistent with the concept of energy and the wave model that they will meet in secondary school.

Skills and attitudes

This is a good topic for developing positive attitudes – the pleasure of music can be linked to scientific understanding at all ages through primary school using appropriate music from nursery rhymes to the pop charts.

Practical work can develop manipulative skills in making sounds, and there are opportunities for simple predictions, e.g. about pitch, loudness and audibility.

Teaching sound

Progression in concepts and contexts

There are a few important concepts to be developed and this can be done progressively with age by the choice of appropriate contexts, e.g.:

Key Stage 1 (Year 1 or 2)

- recognising and naming familiar sounds; encouraging listening with use of stories and sound effects;
- hearing sounds in different places – loud and soft, far and near; 'Can you hear silence?' *Sound is what you have to make or there would be silence.* Robin (5)
- use of familiar and home-made instruments to make and compare sounds.

Key Stage 2 (Year 3 or 4)

- predicting and making different sounds with musical instruments – high and low, short and long;
- looking carefully at what happens to things that make sound. *The cymbal shakes, so it makes a noise.* Rachel (7)
- communicating with sound – making a string telephone and explaining how it works.

Key Stage 2 (Year 5 or 6)

- 'hunt the vibration' in musical instruments and electronic sound makers (e.g. loudspeakers);
- compare a range of instruments from different parts of the world; identify common principles in the way they work, e.g. percussion, string and wind;
- use the idea of a vibration to explain how sound gets to the ear and we then hear it. *The vibration comes off the tuning fork, goes through the air, into your ear and hits the ear drum.* Lisa (9)
- investigate ways to make sound louder and softer using different types of materials; relate this to making music and to noise reduction.

Assessment

The outcomes of the above activities, written, drawn, shown or talked about by the pupils, can be assessed for level according to the descriptors in the National Curriculum:

- Level 1: 'Pupils describe changes in sound. They recognise that sound comes from a variety of sources and name some of these sources.'
- Level 2: 'Pupils compare similar phenomena such as the loudness or pitch of sounds.'
- Level 3: 'Pupils begin to make simple generalisations about physical phenomena, such as explaining that the sounds they hear become fainter the further they are from the source.'
- Level 4: 'Pupils make generalisations about physical phenomena, such as sounds being heard through a variety of materials.'
- Level 5: 'Pupils begin to apply ideas about physical processes to suggest a variety of ways to make changes, such as altering the pitch or loudness of a sound.'

Safety

Children should never put anything into their ears smaller than their fingers. They should be warned of the dangers of prolonged exposure to loud sounds.

Children and light

Knowledge and understanding

The challenge in developing children's ideas about light is that it is so ubiquitous, and vision is so natural, that it is all taken for granted. To many children it is darkness that needs explaining, not light! It seems to have more substance, and the language they meet, e.g. in fairy stories, encourages this. Children will have had little experience of total darkness, and this is not always possible to arrange in school, but even a partial blackout does enable them to see beams of light.

Children may be confused by the use of the word 'light' as an adjective describing colour or weight. They do not readily distinguish between sources that emit light, such as the Sun, and those that reflect light, like the Moon. When children first accept the presence of light, it is not understood to travel. It is 'just there', perhaps around the source as 'a pool of light' or 'so we can see things'. Shadows may be seen as the property of the object, not as anything to do with the light illuminating it. Shadows can also be confused with images or 'reflections'.

A shadow is light that looks a bit like you. It reflects as you move. Jason (5)

Many pupils have a 'person-centred' view of sight; we see things because we look at them, light just helps. An understanding of the concept of light rays is needed to enable pupils to accept that light has to enter the eye. This concept is also needed to explain the formation of images in mirrors and by refraction. This will require further work at Key Stage 3.

Children tend to think that colour is an innate property of an object which light just 'brings out'. Young children think that light is colourless and colour filters add colour rather than subtract it. Experience with prisms and with coloured lights can help challenge this.

Attitudes and skills

Everyone has feelings about light, dark and colour, and these can provide motivation and contexts for the work, e.g. relating it to celebrations, adventure stories or painting. For Key Stage 1 there is opportunity for enjoying the interaction with light and colour in the natural environment, and in decorating models and personal possessions. Key Stage 2 pupils can begin to appreciate the importance of sight to humans and other species and the use made of light in keeping safe, for example reflective clothing for cyclists. They may also speculate on some of the 'big' questions such as 'Why is the sky blue (and red at sunset)?' and 'What is starlight?' (see next section).

The predominant skill is naturally that of visual observation. There is less scope for measuring except for derived quantities such as how far away something is visible. There is scope for prediction, e.g. in the mixing of colours, and for language development and other communication skills.

Teaching light

Practicalities

The paradox of teaching about light is that it really helps to have darkness! In a dimmed room, or using a large box with a viewing window, you can show:

- light beams and rays travelling in straight lines (some dust in the air helps make them visible);
- colour mixing with lights and how it differs from paints;
- coloured light falling on coloured objects;
- reflection by mirrors and refraction by lenses – a light box with slits to let out narrow rays is best for this.

An overhead projector is a powerful light source and useful for showing coloured lights. Optical fibres can show that light travels (even round bends, so make sure pupils have a secure understanding of light's straight line travel first!).

Challenging misconceptions

These are some suggestions researched by the SPACE project (Osborne 1990) and others:

- check children's ability to distinguish primary sources of light (talk about them 'making light') from other things we see, to encourage the idea that light is something to be studied;
- encourage the idea of light travelling in all directions; ask pupils how far away they can see a bright light and why they can see it;
- observe objects in dimmed light, try to make out what they are and think about why we see them – and why we can't in blackout – to challenge the 'active eye' theory;
- present older children with the explanations (of other children) of how we see a book (Figure 4.36), and ask how they would test which one is correct.

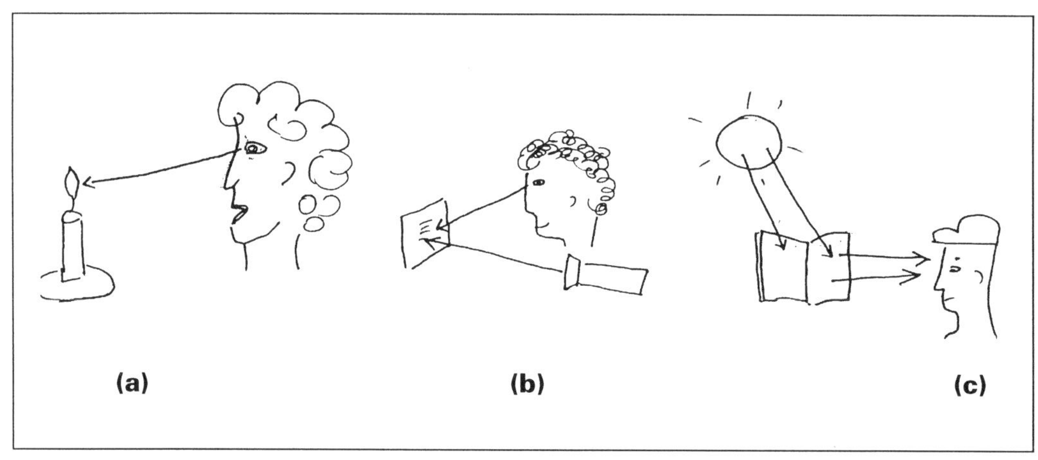

Figure 4.36 Children's drawings of how we see

EARTH AND SPACE

This topic can be daunting in its scope and complexity. For many children (and adults) it is a prompt to the imagination which fills the universe with space travel, aliens and mysterious forces which affect our everyday lives. For others it inspires a sense of awe and wonder related to spiritual experience.

More prosaically, astronomy is a different kind of science from most other topics. It depends heavily on observations, models and theories. The opportunity to carry out 'hands-on' experiments are very limited! Many of the direct observations cannot even be made in school time as the objects are only visible at night. So astronomy calls on a different range of skills than most other areas of science, and makes considerable use of secondary sources of information.

Understanding Earth and space

Time

Our lives are ruled by time, yet we rarely stop to think – what is time? It is cyclic with night and day for example, to which the biorhythms of our bodies are adjusted. It is linear, with the past behind us and the future ahead; and it is always now in the present. It is bound up in the way everything happens – cause always precedes effect. When you run a film backwards you see impossible things like broken crockery reassembling into a whole. Modern physics since Einstein has understood time as a kind of fourth dimension of space-time. All these are aspects of time which may need to be considered in science.

At the primary level, children's awareness of the patterns of time in their lives can lead to a study of how the natural cycles we experience are marked or caused by the motions of the Earth and Moon. The rotation of the Earth on its axis once every 24 hours causes day and night, its annual orbit of the Sun produces the seasons of the year, and the 28-day orbit of the Moon gives us months (Figure 4.37). In these days of air travel and long-distance phone calls or emails it is well

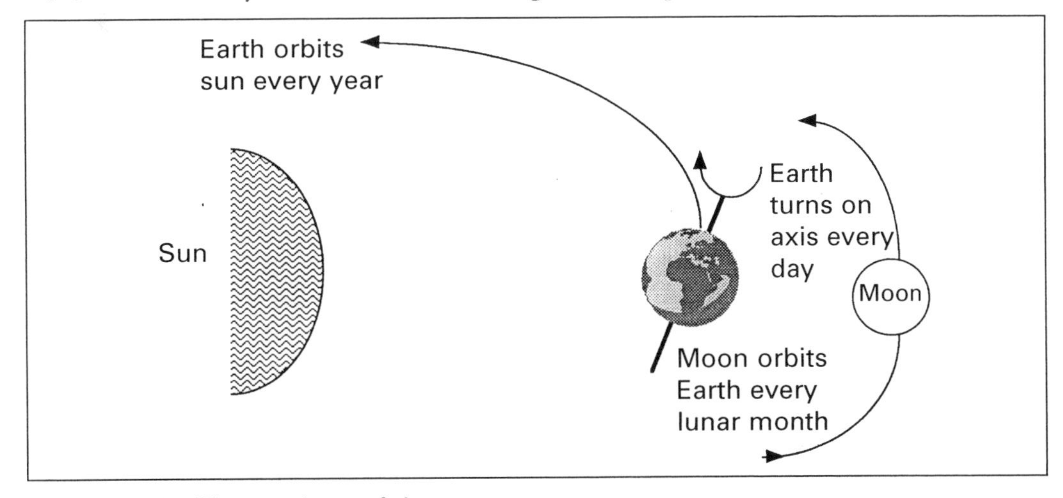

Figure 4.37 The markers of time

known that different places experience different times and seasons due to their positions relative to the Sun and Moon.

Earth, Moon and Sun

From the earliest times, people have tried to explain the nature and movements of what they saw in the sky. These explanations are still developing, but most of what we now know seems contrary to observation and experience, e.g. the Earth orbits the Sun at a speed of about 100,000 km/h, while rotating with a speed at the equator of about 1,700 km/h – and we don't feel a thing!

It is currently believed that the Earth, Sun and Moon were all formed together with the rest of the solar system nearly 5 billion (5×10^9) years ago. The Earth and Moon are made of rocky material (which is molten below the surface) and the Earth has an atmosphere while the Moon does not. Despite their similar appearance in the sky, the Moon is one-sixth the size of Earth while the Sun is 100 times larger. This is because the Sun is about 150 million km from Earth, about 390 times further away from us than the Moon.

The system is held together by gravity, the force of attraction between these massive objects. We can think of the Earth as moving at just the right speed to maintain its distance from the Sun, rather like swinging a mass around on the end of a string. In the same way, to launch a satellite, space engineers have to ensure that it reaches its correct height at the right speed. It will spiral back to Earth if it is too slow, or out into space if it is too fast.

The Moon is a satellite of the Earth, though much larger and further away than any artificial satellites which have been launched. It turns once every orbit, so we always see the same face of the Moon on Earth. During the monthly orbit its position relative to the Sun changes (Figure 4.38). Different amounts of the Moon are illuminated and we see the characteristic cycle of phases from crescent at 'a' to full at 'd' and back to the new moon at 'h'.

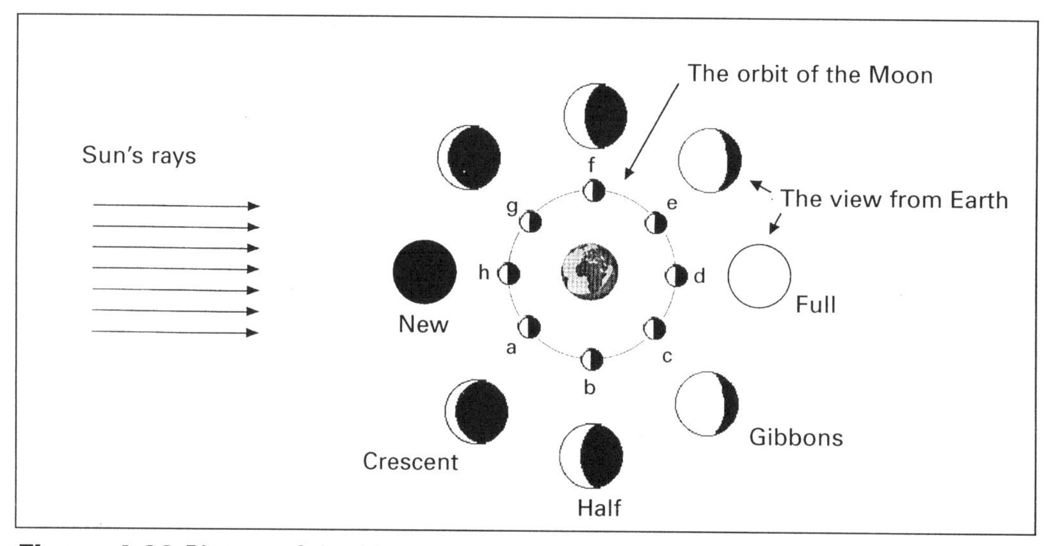

Figure 4.38 Phases of the Moon

The apparent movement of both Sun and Moon in the sky is due to the Earth's rotation. This is in a direction west to east, so time is later in the West and we all see the Sun rise in the East. The seasonal variation is due to the inclination of the Earth's axis. This is at 23.5° to the vertical and it means that part of the Earth is tilted towards the Sun and the opposite side is tilted away. Tilting towards the Sun increases the amount of sunshine. The Sun stays above the horizon for longer and rises higher in the sky, giving the longer, hotter days of summer. Six months later the Earth is on the other side of the Sun and the sides tilted towards and away are reversed. So summer is replaced by winter and vice versa (see Figure 4.39).

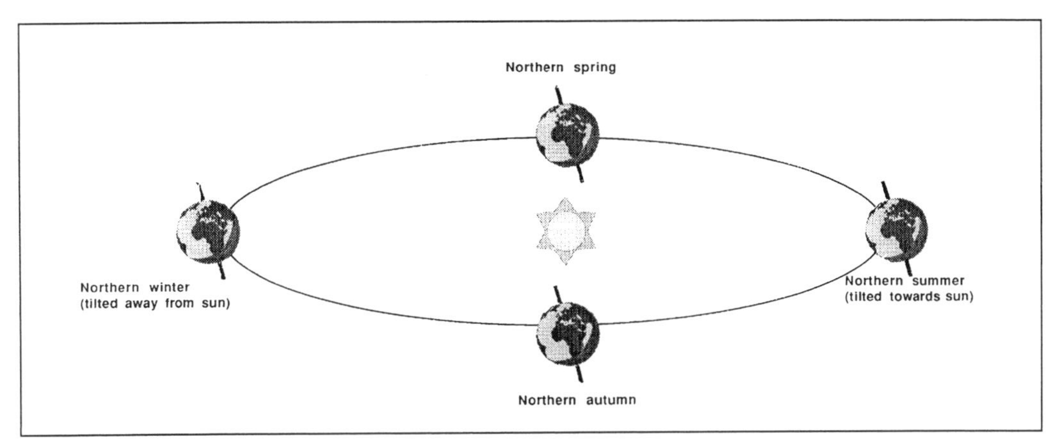

Figure 4.39 The Earth's orbit and seasons

Notice that the effect of the tilt varies, depending on latitude. At the equator there is no difference, so there are no seasons in equatorial regions. By contrast the polar regions receive no sunlight for several months, i.e. 24 hour nights, and have 24 hour days during summer, although it is still cold because the Sun remains low in the sky (Figure 4.40).

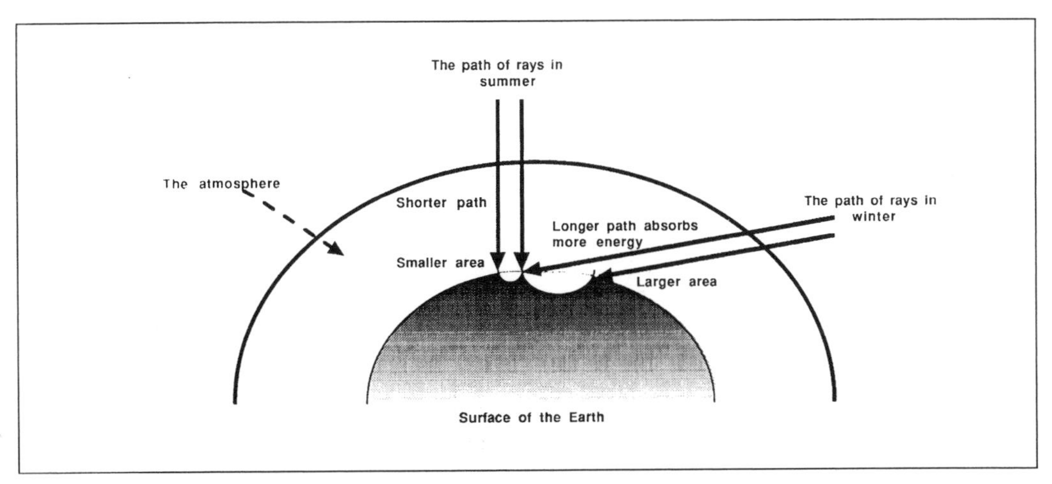

Figure 4.40 How the height of the Sun in the sky affects temperature of the Earth

The solar system, galaxies and the universe

Our solar system consists of the Sun at the centre with nine planets in orbit around it (Figure 4.41). The Sun is a star of average brightness, size and age. It just looks different to us because it is so much nearer to us. The next nearest star, Proxima Centauri, is over a quarter of a million times further away! Stars are made of gas and give out immense amounts of energy from the nuclear reactions which fuel them. In our Sun the fusion reaction which changes hydrogen into helium is about halfway through, so astronomers reckon the Sun will last about another 5 billion years.

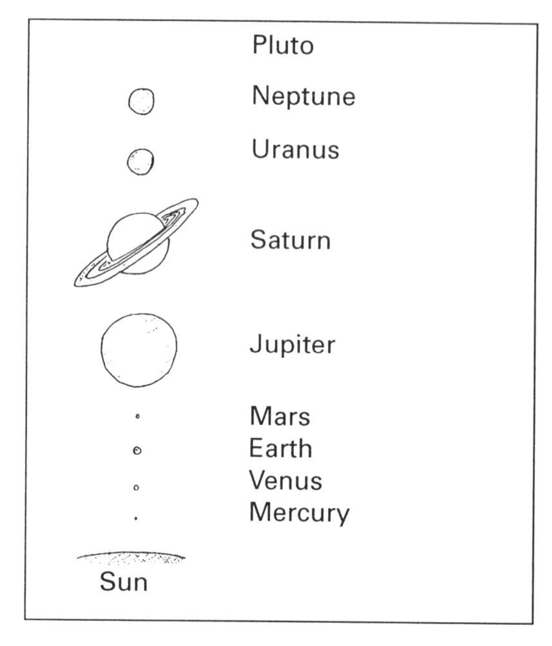

Figure 4.41 The solar system (in order and to scale in size but not distance)

Table 4.5 Data on members of the solar system

Body	Diameter	Mass (compared to Earth)	Surface gravity (compared to Earth)	Density (kg/m³)	Average distance from Sun compared to Earth	Period of orbit (years)	No of moons (*plus rings)
Sun	109.00	333,000.00	28.00	1,400			
Mercury	0.40	0.06	0.40	5,400	0.40	0.20	0
Venus	0.95	0.80	0.90	5,200	0.70	0.60	0
Earth	1.00	1.00	1.00	5,500	1.00	1.00	1
Mars	0.53	0.10	0.40	4,000	1.50	1.90	2
Jupiter	11.18	317.00	2.60	1,300	5.20	11.90	16*
Saturn	9.42	95.00	1.10	700	9.50	29.50	15*
Uranus	3.84	14.50	0.90	1,600	19.20	84.00	5*
Neptune	3.93	17.20	1.20	2,300	30.10	164.80	2
Pluto	0.31	0.0025	0.20	400	39.40	247.70	1

All the inner planets and the asteroids are rocky in composition. The outer planets are made of gaseous substances, but because they are so cold this is often in a liquid or solid form. The outer planets have several satellite moons, often made of rocky materials. The order of the planets can be remembered with the mnemonic: My Very Easy Method Just Shows U Nine Planets. Data on the planets are given in Table 4.5.

The solar system is gradually being explored (a very expensive business!) by rocket-driven spacecraft, sometimes manned by humans, but usually housing a range of instruments. Some of the achievements are listed in Table 4.6.

Table 4.6 Exploring the solar system

Date	*Event (country)*
1957	Sputnik 1 – first satellite launched by USSR
1961	Yuri Gagarin – first man in space (USSR)
1963	Valentina Tereshkova – first woman in space (USSR)
1969	Neil Armstrong – first man on Moon (USA)
1971	Salyut 1 – first orbiting space station (USSR) (USA's Skylab 1973)
1977–89	Voyager 1 and 2 – fly past Jupiter (1980), Saturn (1983), Uranus (1986) and Neptune (1989) (USA)
1984	Challenger – first flight of the reusable space shuttle (USA)
1989	Hubble space telescope launched (international)
1991	Helen Sharman – UK's first astronaut (USSR flight)
1997	Probe lands on surface of Mars (USA)

Exploration of the universe is mainly carried out by land-based telescopes, but these need to be on the tops of mountains because of atmospheric and light pollution. The Hubble telescope is in an even better place, orbiting completely beyond Earth's atmosphere. Countless stars can be seen with the unaided eye, but this is often impossible, especially in towns, because of all the street lights. On a dark night it is possible to see constellations of stars. These are arrays which have been given names to represent pictures, such as The Plough and Orion the hunter. They can be useful for navigation, e.g. the pointer stars point to the pole star; this is always in the north (because it lies in the direction of the Earth's axis of rotation (Figure 4.42)). The stars in a constellation have no actual connection with each other, as they are usually vast distances apart.

When the sky is very dark it is also possible to see the Milky Way, so called because it is composed of so many faint stars that it looks like a trail of spilt milk, stretching across the sky in a narrow band. This is in fact the galaxy to which our solar system belongs. It is immensely large and yet there are billions of galaxies in the universe. Interstellar distances are so great that a special unit is used to measure them. The light year is the distance that light travels (at 300,000 km/s) in a year. This is a distance of almost 10,000,000,000,000(10^{13}) km.

Some sense of the scale of the universe may be gained from an imaginary journey travelling at the speed of light (which physicists think is impossible!) from Earth to the edge of the universe. Table 4.7 shows how long it would take.

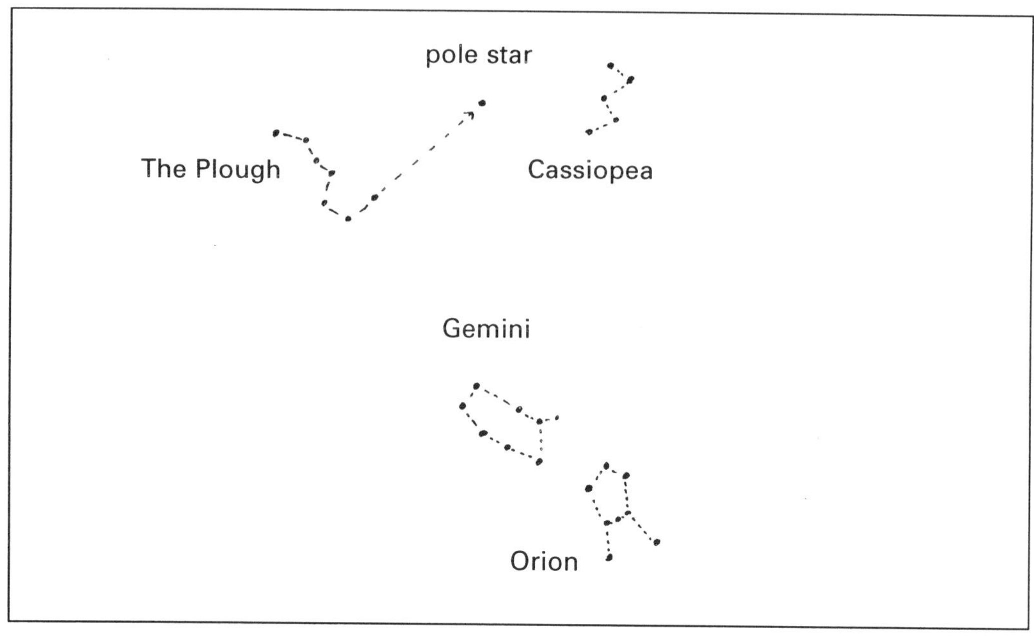

Figure 4.42 Star constellations: the dots represent stars, the dotted lines are added to help identification

Table 4.7 Travelling through the universe at the speed of light

Journey from Earth to:	*Time taken*
The Moon	1.3 seconds
The Sun	8.3 minutes
Pluto (edge of solar system)	5.3 hours
Proxima Centauri (nearest star)	4.3 years
Our neighbouring Andromeda galaxy	2 million years
Farthest star known	15 billion years

Throughout the universe there are cycles of star birth and death. Some stars are so large that when they burn out they collapse with a gravitational force so great that not even light can escape. These are the so-called black holes, whose theoretical investigation by Stephen Hawking has helped us understand the origins of the universe. The generally accepted theory is called the Big Bang for obvious reasons, and physicists are able to describe from the tiniest fraction of a second after it happened how this event led, by expansion, to our present universe. Many important questions still remain to be answered. What is the future of the universe? A 'Big Crunch' is one possibility where the present expansion of the universe is halted and gravity pulls everything back together again. What is beyond the universe in space, or before the Big Bang in time? Can there be space or time, there or then? Can science have anything to say about things that are beyond space and time? Is your brain beginning to hurt yet . . . !

Children and Earth and space

Knowledge and understanding

This topic has been extensively researched by the aptly named SPACE (Science Processes and Concepts Exploration) project (Osborne *et al.* 1994). Many children's ideas were discovered which are at variance with the scientific picture; some of these are shown in Table 4.8. This is hardly surprising since a recent survey of the understanding of science by 2,000 adults found only a third knew that the Earth went round the Sun once a year. What has been mapped is a kind of evolution from the simple intuitive description, usually found in the youngest children, to more complex ideas which show the ability to use models and to imagine looking at things from a different perspective. This perhaps suggests the kind of activities which should engage children and challenge them in primary school science.

Children tend not to differentiate between planets and stars, nor appreciate how far away they are, and to think that the constellations represent connected stars. Many children think of the Earth as being at the centre of everything, and later that the solar system is at the centre of the universe. They have little idea of the dimensions of the universe, or that it is changing.

Children need experience of looking at things which move and seeing views changing while moving if they are to progress. Models and experiences are suggested below. Older children can study some of the history of the topic to see how theories have changed and why.

Children can make regular observations of the position of the Sun, and of the shadows it casts, both during one day and for longer periods. They can then try to associate this with theories of Earth's movement. They will need to use their own time to make regular observations of the Moon and to relate the idea of phases with the Moon's position.

Attitudes and skills

Most children will have met fictional aliens in books, films and on TV. This can be motivating for a study of space, but the fictions will usually conflict with the facts! It is worth encouraging children to look for specific things in the night sky when it is clear and dark; many will not have noticed what you can see when the conditions are good.

The skills of interpreting models and drawings are demanded in much of this work. The teacher needs to be clear what assumptions are made in these, and whether the child recognises these and accepts them. The study will also call for the use of secondary sources of information. Children need guidance on both how to find what they are looking for, and what use to make of it. CD ROMs are

Table 4.8 Progression in children's ideas about Earth and space, as shown by typical representational drawings

Drawing	*Description*
1 Earth in space	
a	 Earth flat, sometimes saucer-shaped
b	 Earth spherical, people live in the upper half
c	 Earth spherical, people live all over but top is up
d	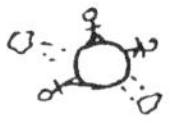 People over all surface, down is towards centre
2 Day and night	
e	 Sun goes down behind hill, or clouds
f	 Moon covers Sun
g	 Sun goes round Earth once a day
h	 Earth goes round Sun once a day
i	 Earth spins/rotates once a day
3 The seasons	
j	 Heavy winter clouds stop heat from Sun
k	 Sun further away in winter
l	 Sun moves to other side of Earth in winter
m	 Earth tilts so parts get winter at different times
4 Phases of the Moon	
n	 Clouds cover part of the Moon; full Moon in summer
o	 Shadow of Sun falls on Moon; no pattern observed
p	 Shadow of Earth falls on Moon; some regularity
q	 Part of Moon illuminated by Sun and part visible on Earth; pattern explained

attractive up-to-date sources which offer the user easy printout and editing facilities. This means children can assemble a most attractive piece of work – without necessarily developing their understanding.

Teaching Earth and space

Practical activities

For younger pupils the topic of time provides lots of opportunities to talk about and record the cycles of everyday life. Different timers can be tried: water and candle clocks, sand timers, rocking tocker timers and mechanical clocks as well as the more common electronic clocks. Pupils can consolidate their grasp of time periods by designing and making a sand or water timer.

Work with shadows and the sundial prepares children for later work on the relative movements of Earth and Sun. Work on the seasons links with environmental studies; the relevance to later work is in noting the regular sequence and features such as the length of days and strength of sunlight (on suitable summer/winter days!). The explanation needs to be left until the upper primary years.

Using secondary sources

The Earth and space is a topic which is well provided for in information books and other media resources, including the newspapers – for older children this can give a topical boost. Pupils need to have a clearly focused task to help them use such resources effectively. This may be gathering answers to specific questions which have been posed by the teacher or raised by themselves. Obtaining data and illustrations of planets will be easier for pupils to handle than explanations of theories and descriptions of relative movements, which will usually need to be supported with class discussions and demonstrations. Presenting the information for display gives another purpose and provides an opportunity to discuss the data (Figure 4.43).

For older children the topic is a good one for raising the issue of the nature and

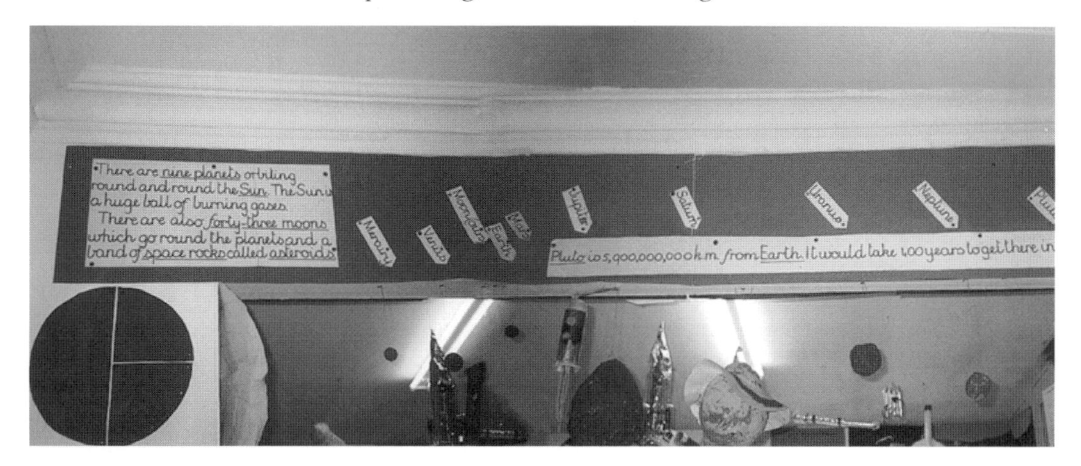

Figure 4.43 Classroom display of the planets

status of scientific knowledge. Statements such as:

- the Earth is a sphere, mostly made of molten rock;
- you can jump higher on the Moon;
- the Sun is a nuclear reactor;
- people in Australia walk upside down;
- the universe started with a big explosion and has been spreading out ever since;

can be used to explore not only pupils' prior knowledge, but also how they know what they know. Do they have any evidence for these things? How would they get evidence that would convince others?

A list of selected resources for children is given in Appendix 2.

Teaching with models

The use of models is an important part of science education. Models enable the learner to understand an idea or theory by relating it to something more familiar and comprehensible. In astronomy, models are essential to enable us to cope with the immense sizes and complex movements and changes which occur. Here is a brief summary of the most useful models for primary schools, fuller details are given in the teaching aids listed in Appendix 2, e.g. *Earth and Beyond* (ASE 1997).

1. A globe Earth (for younger children this could be inflatable, for juniors one which has the inclined axis is essential to explain how the seasons are caused):
   - children can stick model people on it to understand what is meant by up and down in different places;
   - shine a bright torch or projector horizontally on to the globe and look at the shadows of the people on the surface;
   - rotate the globe in the 'sunlight' to show night and day moving around the world;
   - for older children link the time zones and international date line to lines of longitude – the Earth turns 15° every hour;

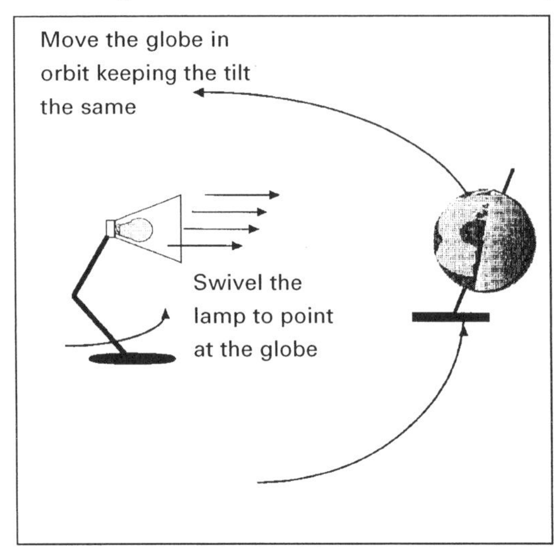

Figure 4.44 Model for showing the seasons

- move the globe in an orbit around the light source, without changing the inclination of the axis (Figure 4.44); compare the tilt and the light patches at opposite sides;
- rotate the globe and look at the patches of light and shadow near the poles, which show where the Sun never sets in the summer and never rises in the winter.

2. A white ball Moon: a large polystyrene ball or similar can be mounted on a stick,

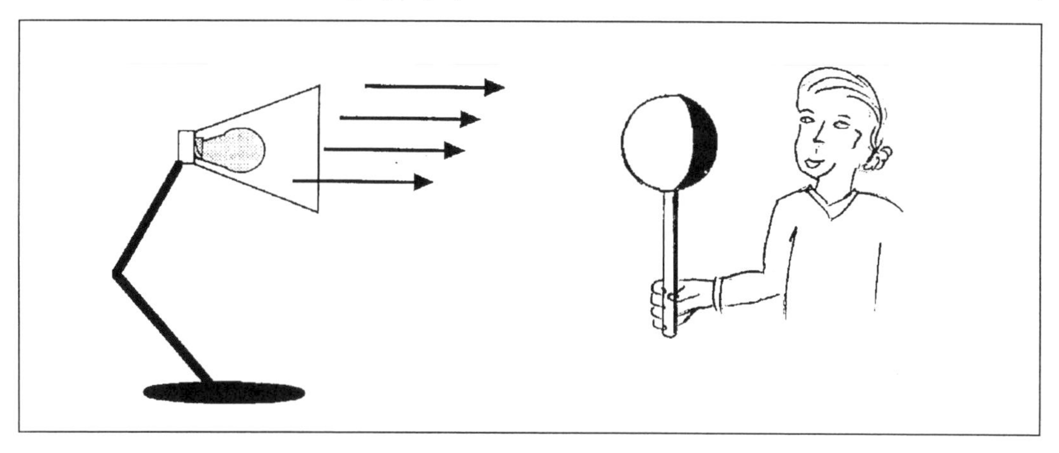

Figure 4.45 Model for showing phases of the Moon

e.g. a knitting needle, so that children can hold it in front of themselves. A strong light is shone on to it, representing the Sun. They become the person on Earth and turn slowly with the Moon to represent its orbit (Figure 4.45). As they look at the Moon they will see the change in what is illuminated – the cycle of the phases.

3. A large (the Earth) and a small (the Moon) ball each mounted on sticks can be

Table 4.9 Scale model of the solar system

Planet	*Diameter (mm)*	*Distance from Sun (m)*
The Sun is a disc 0.5 m in diameter, which is a scale of 3,000 million to one		
Mercury	2	15
Venus	5	30
Earth	5	50
Moon	2	(0.1 from Earth)
Mars	2.5	70
Jupiter	60	250
Saturn	50	500
Uranus	20	1,000
Neptune	20	1,500
Pluto	2.5	2,000

3. A large (the Earth) and a small (the Moon) ball each mounted on sticks can be used with a torch (representing the Sun) to explore eclipses. A lunar eclipse is when the Earth's shadow falls on the Moon. A solar eclipse is when the Moon's shadow falls on the Earth.
4. An electrically or hand-driven model of the solar system (called an orrery) shows the relative movements of the planets around the Sun. The planets can be drawn, coloured and cut out. It is good to do this to scale and to discuss with pupils how to display them in their positions in the solar system to scale also. Table 4.9 shows a common scale for both planet sizes and distances. It is hardly a practical one for display with Pluto 2 km away, but it does show that most of the solar system is empty space!

Assessing what children think

A popular question for exploring what children think about space is to ask them to imagine they are in a spaceship and to draw what they would see out of the window – the question has even appeared on a SAT paper! Alex Lundie's class of Year 2 pupils were able to show they knew that the Sun and Moon were separate spherical bodies, but in discussion he found that they were unable to use this knowledge in their explanations of everyday experience such as day and night (Lundie 1994). Extensive research with secondary pupils in the USA has shown that it is very easy for teachers to assume that pupils have a fuller understanding of this topic than they do (BBC Education 1994). We need to remember how complex and remote most of the ideas of astronomy are, and appreciate that children's own ideas, based on their everyday experience are remarkably resilient. Try completing the concept map in Figure 4.46 to check your understanding and think how you might use a similar one with a class.

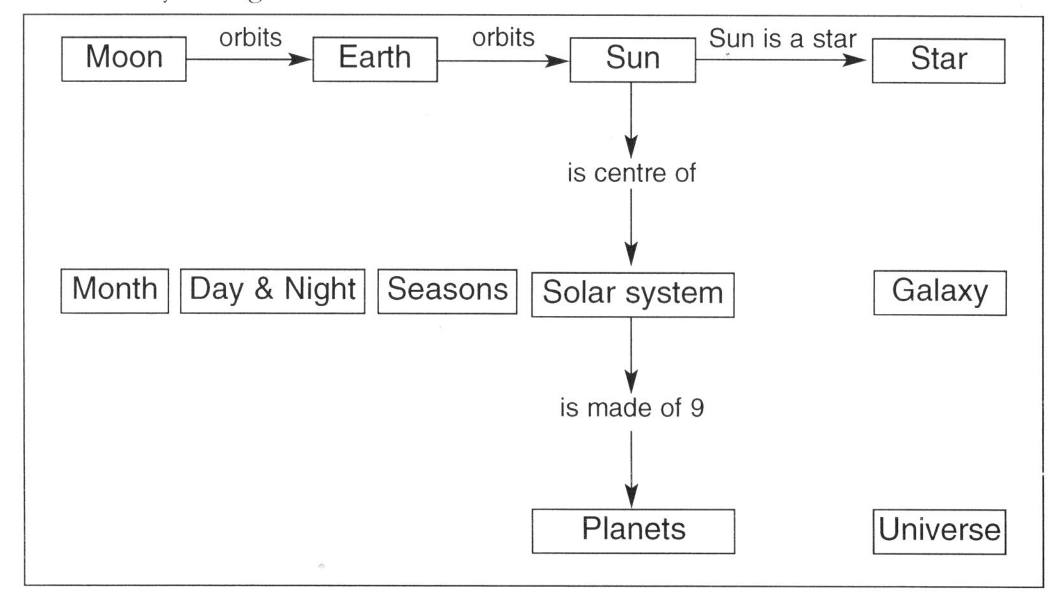

Figure 4.46 Concept map of Earth and space

CHAPTER 5

Planning and Assessing in Primary Science

Planning and assessing science are the main focus of this chapter. They are both fundamental to effective teaching and learning in science. Assessment must inform planning and therefore must be considered as an integral part of the planning process. This chapter examines the issues which underpin planning and assessment, in addition we consider science within the broader context of the whole curriculum.

PLANNING

Planning needs to address the requirements of the National Curriculum for Key Stages 1 and 2, and the Desirable Learning Outcomes for nursery and early years units. In addition to meeting these requirements, planning must also provide progression and continuity for children throughout the primary school. It is obviously essential to consider how the concepts and activities taught build on previous work, with consideration of how it can make meaningful links with other areas.

What is meant by a scheme of work?

There has been a vast improvement in the teaching of the science curriculum in primary schools during the last decade. Schools have now firmly established science as an integral part of the curriculum, although for many schools there are still concerns, particularly in relation to effective planning and organisation of the science curriculum. These have also been clearly identified through many OFSTED reports:

> The major issue in primary science is the effective planning of science within the topic based structure of most primary teaching schemes. With the exception of such areas as curricular planning and the need to improve teachers' subject knowledge, primary science can be counted as a success. (OFSTED 1994)

Schools have used a variety of terminology over the years to describe aspects of the science policy and/or curriculum. Common terms used are policy, guidelines, programme of work, curriculum map, whole school plan, scheme of work, long-term, medium-term and short-term planning. Each of these has a slightly different focus, and therefore can be used to provide guidance and information for a variety of audiences, e.g. teachers, non-teaching staff, parents and governors. There is no clear guidance as to what a policy or scheme of work should look like, and a certain amount of confusion therefore surrounds the range of terms. However, the Association for Science Education (1996) produced a leaflet, *How to write and use a scheme of work*, to explain what is meant by the variety of different titles. It suggests the format in Table 5.1 is a useful starting point for streamlining the varied interpretations.

Table 5.1 Science terminology

Aspect of consideration	*Could be called*
What is science in our school all about?	Science policy
What are the important principles on which we base our science teaching?	Science guidelines
Which parts of the National Curriculum should each teacher aim to teach to their class?	Scheme of work informing long-term planning
How will the curriculum be divided up?	Curriculum map informing whole-school progression
What is it I want my children to learn?	Scheme of work
How will I find out where they are?	Medium- short-term planning
How will I get them there?	Termly/half-termly plans
What activities will we do?	Appropriate record keeping and assessment

It is essential for a school to have a well-planned, whole-school scheme of work which clearly identifies the point at which the children are first introduced to a scientific concept, and when and how they are to revisit the concept. By working to such schemes teachers will be able to monitor children's learning in science and therefore provide continuity and progression. It is, however, important to remember that evaluation should be built into the scheme. Once a scheme is written it should be considered an ongoing working document. By constantly reviewing and evaluating the scheme of work schools are ensuring that science teaching and learning are progressing.

The majority of schools now adopt a curriculum map across the early years, Key Stages 1 and 2, to show when children are introduced to a science concept and when these are revisited. It is essential that each year children are covering aspects of the National Curriculum for science or the appropriate Desirable Learning Outcomes. Science 2, 3, and 4 are the main areas of experience in terms of knowledge and understanding, with Science 1 being developed throughout the knowledge and content areas. As can be seen from the curriculum map (Table 5.2) the children are engaged in aspects of life processes and living things (Science 2), materials and their properties (Science 3) and physical processes (Science 4) each year. These will obviously have a different focus depending on the age of the children and the previous experience they have of each of these areas. What is essential is that teachers are clear as to how and when experimental and

Table 5.2 Curriculum map

	Reception	Year 1	Year 2	Year 3	Year 4	Year 5	Year 6
Autumn term	My family	Sound and music Sc4	Materials – homes Sc3	Earth in space Sc4	Built environment Sc2	Weather and its effects Sc3	Forces Sc4
	Day and night	Healthy eating Sc2	Magnets Sc4	Living processes Sc2	Hot and cold Sc3	The solar system Sc4	Our environment Sc2
Spring term	Connections	Materials – clothes Sc3	Growth Sc2	Electricity Sc4	Materials Sc3	Human body Sc2	Materials Sc3
	Growing plants	Floating and sinking Sc4	Keeping healthy Sc2	Uses of materials Sc3	Frictional forces Sc4	Food chains Sc2	Electricity Sc4
Summer term	Everyday objects	Local habitats Sc2	Moving things Sc4	Light–shadows Sc4	Habitats Sc2	Light sources Sc4	SATS
	The pond	Rocks, soil and weather Sc3	Using energy Sc4	Variety of life Sc2	Sound and music Sc4	Materials – properties Sc3	Sex education Sc2

investigative science (Science 1) is drawn into each aspect of science as it is taught.

This method of mapping out the science concepts avoids fruitless repetition for children, and the teachers know exactly what the children have experienced previously. It is important to note here that we are not implying that children do not need to revisit areas. It is essential that they build on these previous experiences. For example, electricity with Year 1 children may involve a battery, bulb and wire. By the time children are in Year 6 they will be designing much more complex circuits, possibly in a problem-solving context, which perform particular functions. Using a curriculum map with previous records and profiles of individual children, teachers should have the necessary information to plan appropriately when revisiting a concept at a later stage. We will refer to records and profiles in more depth later in this chapter.

The main aim of a scheme of work should be to indicate a strategy for providing a broad and balanced curriculum. As previously stated however, there is no prescribed format for a scheme of work, but it will include progressive learning objectives, contexts, activities and outcomes and therefore detail most of the following issues:

- planning;
- organisation of the children;
- organisation of materials and resources;
- methods of assessment and monitoring.

The headings may vary depending on the way in which a school organises its schemes of work across the whole curriculum.

Long-, medium- and short-term planning

In the document *Planning the Curriculum at Key Stages 1 and 2*, SCAA (1995) identifies three broad levels of planning, each of which has its own particular purposes and outcomes:

- **Long-term planning** is concerned with producing a broad curriculum framework for each year of the key stage. It reflects the school's overall curricular aims and policies, and the whole staff and governors are involved at different stages of the process.
- **Medium-term planning** deals with the details of the scheme of work to be taught to each year group and identifies opportunities for assessment. It involves year group or key stage teachers often supported by coordinators.
- **Short-term planning** is usually carried out individually by class teachers and is used to focus day-to-day teaching and assessment.

What is clear is that the medium-term planning is the most essential component in terms of teaching and learning, not just for science but for all aspects of the curriculum. Although the levels of planning set out by SCAA (1995) are most commonly adopted by schools, no government agency has issued an exact model of what a good medium-term plan for primary science might look like.

Table 5.3 Medium-term planning for electricity

Key questions	*Skills/concepts*	*Activity*	*Resources*	*Assessment*	*Differentiation*	*Progression*	*Teaching strategies*	*Cross-curricular links*
Can you think of anything which needs electricity to work? What can you tell me about electricity?	Electricity can come from power-stations and batteries. Mains electricity can be dangerous. AT1, 2, 3 & 4 Possible health and safety.	Discussion of children's ideas of electricity. Health and safety. Brainstorming. Draw ideas. Write sentence.	Examples of electrical appliances and battery operated items i.e. torch, car.	Did children offer ideas? Ability to recognise mains operated appliances. Ability to recognise dangers.	Allow less confident to participate. Encourage & give praise for correct and wrong answers. Help with writing sentences.	Clarification of personal ideas and . development.	Whole class seated in story area	Art English
What things use electricity? Where would you find them?	Electricity is used in different ways: heating, lighting, making things work. AT4 LEVEL 1	Classifying battery & mains operated objects. Drawing, writing. Graph findings.	Collection of objects which use electricity.	Able to sort mains and battery operated objects including toys.	Pictures of objects to cut out and apply in a graph (less able). Graph on computer (middle group) Own graph and drawing (able).	Knowledge from previous activity.	Whole group revisit main points with less able.	Maths English Art History
What is the difference between these batteries? How can you make a bulb/toy brighter/faster?	Batteries come in various sizes and strengths They are safe and convenient to use. AT4 LEVEL 2	Using batteries to make a bulb glow bright. Toys to make them go fast/slow. Measuring distance travelled by toy. Fair testing. Write up findings.	Selection of various sized batteries and voltages. Run down batteries. Toys and bulbs various wires.	Able to distinguish size and voltages. Able to use appropriate battery for purpose.	Assist less able to write sentences. Extend activity for more able. Simplify questions or use others to promote enquiry.	Knowledge of Health and Safety gained from previous activity. Electricity can be stored.	Groups of same ability. Teacher to spend more time with less able.	Maths English
How can you make a bulb light up? Can you draw how you achieved it?	There needs to be a circuit for electricity to flow. Some materials allow electricity to flow through them. AT4 LEVEL 2	Investigating; hypothesising; experimenting; diagrams of circuits; Sorting conductors & non-conductors.	Battery, bulb, low voltage flex. Selection of conductors/non-conductive materials. Professors ' progress game contains these.	Were children successful? Able to identify materials that conduct electricity?	Groups of similar ability peers assist after a period of time. Progressive questions. Revisit previous related activities.	Knowledge of previous activity. Progressed from using bulb holder to exposed bulb and flex.	Groups of 6 children. Whole class discussion of previous activity then group work.	Maths Art English
How can you make a buzzer sound? How will you make it stop and start?	Switches are convenient ways of completing and breaking circuits. AT4 LEVEL 3	Circuits: making and breaking as part of topic building a room with lighting and door bell/buzzer.	As above, but incorporate buzzer and paper-clip switch.	Able to make/break circuit to sound buzzer, light bulb. Use appropriate science vocabulary.	As above previous activity.	Knowledge of previous activity. Develop, look and investigate magnets.	Groups of same ability activity. Whole class discussion.	D&T Maths English Geography

Many schools have now adopted the chart format for medium-term curriculum planning like the one set out in Table 5.3. These provide the teachers and head teacher with a coherent overview as to exactly what the children will be doing in science throughout the school. This helps teachers to be, 'clearer about the learning objectives and the learning pathways that characterise pupils' progress in science' (OFSTED 1996). The detailed medium- and short-term plans teachers have developed provide clear criteria which can be used to assess the children's scientific learning.

Planning for differentiation

Differentiation means meeting the different learning needs of the individual children. Differentiation is not something that can be tagged on to a scheme of work; it must be an integral part of the planning stage (see medium-term planning for electricity, Table 5.3). Whenever teachers set children a task or activity, they monitor how the children respond and cope with the task. When it seems appropriate, we intervene either to help the child or to stretch the child further. Differentiation is therefore seen as a process of identifying the most effective strategies for each learner. We can provide differentiation in various different ways through the resources used, tasks set and outcomes.

Planning science for the early years

Although with older children science may sometimes be taught as a separate subject, in an early years setting it needs to be considered within the holistic nature of child development.

The Desirable Learning Outcomes (SCAA 1997) now provide a framework for looking at children's learning before they enter school. There are six areas which have been identified:

- personal and social development;
- mathematics;
- language and literacy;
- knowledge and understanding of the world;
- physical development;
- creative development.

The area which encompasses science is knowledge and understanding of the world, which:

> Children talk about where they live, their environment, their families and past and present events in their own lives. They explore and recognise features of living things, objects and events in the natural and made world and look closely at similarities, differences, patterns and change. They show an awareness of the purposes of some features of the area in which they live. They talk about their observations, sometimes recording them and ask questions to gain information about why things happen and how things work. They explore and select

> materials and equipment and use skills such as cutting, joining, folding and building for a variety or purposes. They use technology, where appropriate, to support their learning. (SCAA 1997)

These outcomes are very broad and allow teachers to focus on a wide variety of children's learning through play contexts. The aim of the Desirable Learning Outcomes is to help teachers to plan an effective curriculum for all 4- and 5-year-olds.

Many practitioners may not feel confident about what comprises science for 3- to 5-year-olds. One of the most useful ways of dealing with this is to analyse the possible 'scientific content' that is contained in the provision made for the children, e.g. using batteries and bulbs not circuit boards. This in turn helps the adult to support the children's learning with specifically appropriate language in meaningful contexts. In this way, scientific concepts are built up via a range of hands-on experiences, which is the most appropriate way for young children to learn. Some examples of these early years experiences have been given in previous chapters.

Developing positive attitudes to science is very important at this stage. No-fail situations where there is not a 'right' or a 'wrong' answer (e.g. if a child is constantly being told they have the wrong answer, they cannot achieve at their own level) helps to build a child's confidence and self-esteem. Once a child has developed a positive attitude to science they can move through their education with confidence and an enquiring mind wanting to find out how and why things work as they do. A self-servicing classroom where the children have easy access to equipment helps to support scientific learning. We will focus further on resources later in this chapter. Also open-ended questions allow children to respond at their own level. A concept is built up from a range of experiences, at the nursery stage therefore one 'scientific experience' should not be used to assess whether a child has gained a particular concept. Rather, recorded observations over time will yield this information. We will refer to observations later in this chapter. In the nursery context, science is most effectively approached through areas of provision and experience, for example, water play, sand play, connecting objects, pouring and filling, pushing and pulling carts and trucks. Science is very rich in the nursery environment, and much of what young children are engaged in as they spend time exploring and investigating are early scientific experiences.

Planning using published schemes

Many schools use published schemes to support and enhance the ideas and views held by the class teachers. However, published schemes vary greatly in the type of support they offer teachers: some schemes provide background scientific knowledge; some particularly highlight the importance of experimental and investigative science SC1, and show how it can be inter-related with the knowledge and content areas, life processes and living things SC2, materials and their properties SC3, physical processes SC4; some schemes identify where and when assessment of the science concepts should take place. It is essential for schools to decide what they want a published scheme to provide. Drawing up a list of criteria, staff should ensure the scheme will meet the needs of the school. The published scheme could also be used to contribute to the school's own scheme of work for science. It may be that it will provide a framework on which the school can develop its own scheme of work.

Organisation of the children

What is the most effective way to organise the children for science activities in the classroom? The teacher will need to decide on 'fitness for purpose'. The organisation of the children will depend on a number of factors such as whether it is an introductory session, a session in which the teacher wants to make observations of particular children, and the scientific concept being taught. Groupings of children for science need to consider the activity in which the children are engaged, and the equipment being used, with particular reference to safety issues. The teacher will also need to consider the appropriateness of children working on their own, and whether an adult needs to be working with the group, or supervising more than one group. A variety of methods for organising science in the primary classroom are set out in Table 5.4.

Organisation of materials and resources

It is important to remember that the greatest resource a school has is the staff who work within it. The sharing of ideas and information is an excellent way to support each other, enabling less confident colleagues to discover the rewards and joy of teaching science. Resources concern more than just equipment: they include the staff, the space, books and schemes used by both children and teachers, as well.

It is essential to have appropriate resources to develop children's learning in science. The organisation of resources is of paramount importance if science is to be taught effectively. Children must have access to appropriate resources and not spend time waiting for resources to become available. Encouraging children to select and use appropriate resources is a skill which they need to develop. The suitability of the resources children select can be an indication of their scientific understanding.

For learning in science to be effective, teachers must be aware of the resources in the school, where and how they are stored, as well as how they can be used. There are clear advantages to having self-servicing classrooms where the children can access basic everyday resources for themselves, not only for science but for all areas of the curriculum.

It is essential that a whole-school policy is agreed, meeting the needs of both the teachers and children. This will necessitate whole-school discussions to work through the issues and problems which are pertinent to each individual school.

The following issues will need to be addressed:

- Where do the children carry out science work – classrooms, science room, wildlife areas, etc?

Table 5.4 Possible ways of organising the children for primary science teaching

Method of organisation	*Advantages*	*Limitations*	*Science concept and skills*
Whole-class teaching for demonstration	Minimum organisational demands Economical on time and equipment	No first-hand experience for the children No differentiation Difficult to involve the whole class	Earth and space – demonstrating a model – or other instances where models are used
Whole-class introduction session	Children can brainstorm their ideas; also learn and develop ideas from each other	Requires effective management of the children so they all have an opportunity to express their ideas	Any concept or topic
Whole-class practical	Children can share ideas	Requires very skilled teacher Maximum organisational demands Can require a lot of resources	Living things – environmental aspects, communication and health and safety
Class practical Children work in small groups doing similar tasks	Children can work at their own pace (extension work must be available) First-hand experience for the children Equipment demands known in advance	Follow-up lines of enquiry difficult Large quantity and duplication of resources Involves much clearing away	Forces and motion Prediction and fair testing
Thematic approach Small groups working independently to contribute to whole class	High interest and motivation Pupils work at own pace Builds confidence in communication skills when reporting back	Difficult to ensure coherence and individual understanding from report back	Materials – sorting and classifying
Circus of experiments Small groups rotating around prescribed activities	High interest and motivation Easy to plan ahead, less demanding on equipment	Pressure on completion time before change over Method of briefing the children is essential	Life processes – senses, observation and investigation
Small groups or individuals	Children work at own pace and to own potential High interest and motivation	Demanding on teacher time Structured framework essential	Electricity – hypothesising and recording

- Which published materials are used, what is their purpose: ideas for background understanding; for appropriate activities; for organisation; for assessment and recording?
- What published materials are used by the children?
- What are the safety aspects and dangers concerned with science activities? How will the staff manage these?
- How well are the resources used?
- Which resources are not used and why?
- Which resources are available in the classrooms? Are they sufficient?
- What resources are shared? What resources are easily shared? What resources cause difficulties?
- Where are shared resources kept? What is the storage system? Are they easy to access? How well are they maintained?
- Who is responsible for maintaining resources?

Schools have to find the best organisation for their particular circumstances which will in addition to the questions already raised inevitably have to take account of:

- the geography of the building;
- the experience and expertise of the staff in teaching science.

All of these issues need to be addressed and organised for resources to be used effectively. As a staff it is important to gain overall agreement. Set out below are some general ideas for resource organisation:

- **Centrally stored resources:** Clearly labelled, centrally stored resources can be made accessible for all staff. They need to be maintained regularly to ensure that consumables and breakages are replaced. This is a very time-consuming job, which frequently falls to the science coordinator. Problems can arise if resources are not maintained and are left in an unsatisfactory condition. If teachers cannot access the equipment they need, or find it damaged, they may not know how to use alternative equipment. There can be problems of tracking where resources are in the school, therefore centrally stored resources need a system set in place to ensure that resources can be traced at all times, such as staff replacing a box taken with a coloured brick that denotes they have the equipment, or signing it out in a book.
- **Class-based or year-group-based resources:** The advantage of having resources in each class room is that they are available as questions and enquiries arise. However, financial considerations will mean they are limited and only perhaps enough for one or two children to use at any one time. This can then pose problems if a teacher wants to work with a whole class or even more than one group. Although year-group resources can facilitate the main topics to be covered by that year, there are frequently problems with more general equipment like magnifiers, mirrors, timers, thermometers, prisms and microscopes.

- **Topic boxes:** These can be helpful to teachers who are unsure of what resources they may most effectively use. However, as with class-based resources, general equipment can be problematic.

We have included in Appendix 3 a list of essential and desirable equipment for Key Stages 1 and 2.

ASSESSING AND RECORDING

Planning and assessment need to be seen as a continuous cycle, as illustrated in Figure 5.1. Assessment needs to be considered at the planning stage so that the teachers are clear about what they are looking for in order to assess the children. Assessment must inform our teaching; so it needs to be formative as well as summative and evaluative.

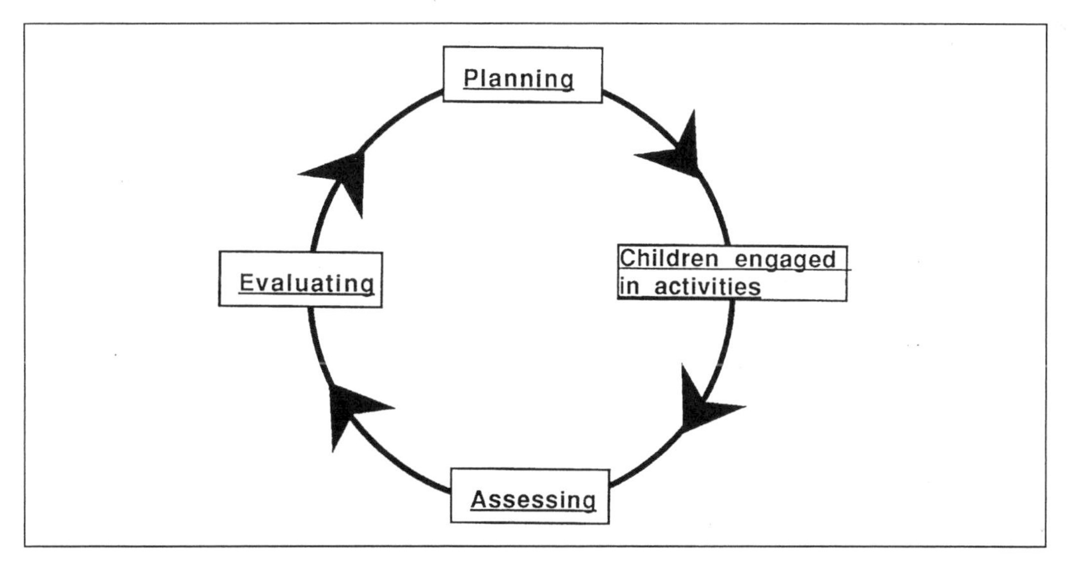

Figure 5.1 The assessment cycle

It may be useful at this point to consider the different purposes of assessment:

- **Formative assessment** is part of the teaching and learning process. Information is gathered regularly as a result of normal teaching and is therefore used for making decisions in ongoing work for the children. It allows the teacher to adjust the activities given to the children to ensure appropriate match. It is usually informal and the children are unaware that it is taking place.
- **Diagnostic assessment** has a more specific focus, being concerned with examining a particular area of performance. It is sometimes assessed through special tests.
- **Summative assessment** involves a summing up of where a child has reached at the end of a particular time. Frequently information is obtained by testing, e.g. key stage standard assessment tasks. Summative assessment provides a picture of a child's development at the end of a year, term, or key stage.

Developing assessment strategies

Although the purpose of assessment is to gain knowledge of the children's understanding, there are differences in the ways assessment is gathered across the curriculum. For example, assessment of reading would tend to be done on an individual basis. This would not necessarily be the case for a science activity as the teacher may work with a small group to assess a particular concept or skill. A variety of strategies can be employed to assess children's understanding of science and these will be considered next.

A useful starting point is to plan an initial activity which provides the children with an opportunity to show you what they already know. Children hold many ideas and views about how the world around them works. It is therefore essential that we allow children to share and explore their own ideas with us. This approach to teaching and learning in science is endorsed through recent research projects such as the primary SPACE (Science Process and Concept Exploration) project (1991). Elicitation is of the utmost importance at the outset of a topic. From this the teacher can build on the ideas the children hold and encourage them to investigate their own ideas.

Useful strategies which will allow teachers to elicit the understanding that children hold about different aspects of science were developed by the SPACE project, and this way of working has also been adopted by the Nuffield Primary Science (1995) scheme. There are a variety of ideas, some of which will be more appropriate than others depending on the area of science you are covering and the age range being taught:

- **Using log-books (free writing/drawing):** Where the concept area involves possible long-term changes, it is suggested that children should make regular observations with the frequency of these observations depending on the rate of change. The log-books could be pictorial or written, depending on the age of the children involved, and entries could be supplemented by teachers' comments if the children's thoughts need explaining more fully. The main purpose of the log-book is to focus attention on the activities and to provide an informal record of the children's observations and ideas.
- **Structured writing/annotated drawing:** The SPACE team found writing or drawings produced in response to a particular question extremely informative. Drawings and diagrams were particularly revealing when children added their own words to them. Annotations can help to clarify the ideas that a drawing represents.
- **Completing a picture:** Children were asked to add the relevant points to a picture. This technique ensured that children answered the questions posed. The structured drawings provide valuable opportunities for teachers to talk to individual children and to build up a picture of each child's understanding.
- **Individual discussion:** It is essential that teachers use an open-ended questioning style with their children. The value of listening to what children say, and of respecting their responses, is important in clarifying the meaning of words children use.

Developing assessment opportunities

Having allowed the children the time to develop and explore their own ideas, it is essential that teachers have already considered ways of monitoring the children to see how effective our teaching has been. As we have already stated, without effective assessment teachers are not able to make appropriate decisions about the next step of the children's learning. Below are some strategies to enable the teacher to find out what ideas have changed, what ideas have remained the same, and what further understanding children have gained of the scientific concept they are working on:

- Talk the children through what they are doing. This provides both the opportunity for children to explore their own ideas, while giving the teacher insights into the children's thinking.
- Listen to and tape a group of children carrying out an investigation and focus on their discussion as they work through it. This again will provide the teacher with useful insights into their understanding.
- Allow the children to report their findings to the whole group (refer to the later section on children's own recording of their science). This provides the chance for children to practise communicating their findings as well as to explain their ideas and reasoning to each other.
- Ask the children to record an investigation in their own way. What they choose to focus on or ignore will again provide the teacher with a useful insight to their level of understanding of the topic.

All of these strategies will help the teacher to build up a knowledge of the children's scientific understanding.

Concept mapping

We have referred to concept maps in the introduction and in previous chapters of this book. The advantages of concept maps are that they are formative and can be completed quickly. The teacher does not need to spend time preparing an assessment task and it does not put children in a test situation. Children's concept maps reflect their own ideas and understanding, so cannot therefore be marked wrong or right, even if their ideas do not match with what is regarded as correct scientifically. Their purpose is for the teacher to make appropriate provision for the children's learning to further develop their ideas and understanding. A concept map can be useful at the outset of a topic to show what scientific understanding children hold. A further concept map at the end of the topic will show the teacher how and in what ways their understanding has developed, and what aspects of the concept still require further work. It may be that the way in which the teacher presents the concept does not match the learning style of the children, so a final concept map will also enable the teacher to evaluate the effectiveness of their teaching.

Developing manageable recording systems

It is essential that close record keeping of children's progress and achievement is kept throughout. Without such records it is not possible to decide on the most appropriate way for developing individual children's learning in science and other aspects of the curriculum.

Records should only be kept of significant progress by each child where they are likely to serve one or more of the following purposes:

- informing future planning;
- informing reports to parents;
- informing future teachers;
- providing evidence for teacher assessment at the end of the key stage.

Records should be useful, manageable and easy to interpret. Schools have developed their own methods and ways of setting up record-keeping systems. These records can collect varied information, e.g.:

- *an assessment focus* on a particular activity can provide the teacher with clear information: if possible outcomes are set up at the outset of the activity the teacher is focused on exactly what he or she is looking for;
- *small group observations* can allow the teacher to make observations on a particular activity and gain an insight into the children's understanding;
- *an observation/task sheet* allows the teacher to focus on particular objectives within a task.

Assessment floating & Sinking Year3 Date 18.4.97
Focus/Activity prediction/sorting

Possible Outcomes:
(1) Sort by shape
(2) sort by material.
(3) sort according to weight - how heavy it feels
(4) relationship between size & how heavy it feels
(5) accuracy of predicition.

Child's Name	1	2	3	4	5	Comments	Further Comments/Needs
Lucy	✓	✓	✓	✓	✓	very confident has a good understanding.	extension work to further her ideas.
Joe	✓	O	✓	O	O	unable to articulate his understanding.	more opportunities to explore his ideas.
Helen	✓	✓	-	-	-	reluctant to predict.	needs to develop confidence.
Dilip	✓	✓	✓	✓	✓	very confident able to make accurate predictions.	extension work to develop his ideas.
Robin	✓	✓	-	-	-	did not focus on size at all.	more time to develop his ideas.

Small group observations Date: 20.1.97 Year 1
Activity: Introduction to electricity - wire bulb battery

Name	Observation	Action
Zanab	says it should fit together - but does not know how.	more time to work with equipment.
Rose	no idea they fit together - "it needs to be plugged in".	lot of practical work.
James	tried to put one end of wire to bulb & one to battery	further practical work
Alex	Copied James - no strategies of his own.	time to explore his own ideas - Confidence
Josie	managed to get the bulb to light.	further extension circuits.

Figure 5.2 Examples of records

What is essential is that the records are meaningful. Some examples are shown in Figure 5.2.

Children's own recording of their science activities

Children's own recording of their science activities can provide the teacher with an important insight into each child's level of understanding. When discussing assessment strategies, we previously suggested a variety of techniques for eliciting children's ideas. These same techniques could be used to provide the teacher with a record of children's progress and development.

In the current National Curriculum (1995), the former statements of attainment have been replaced by level descriptors. It is intended that teachers will refer to these when summarising a child's achievements at the end of a key stage, rather than using them as a basis for the day-to-day assessment of a child's progress. The level descriptors indicate the types and ranges of achievement that are characteristic of that level. Teachers are asked to use their knowledge of a child's work to judge which level descriptor best fits the child's achievement when the summative assessment is being made at the end of a key stage.

It is therefore essential that teachers keep ongoing records of children's achievements on a regular basis so that they can make accurate assessments of them from an informed basis of evidence gathered throughout the year.

Observing and assessing for the early years

It has already been noted that the nursery environment provides a rich source for early scientific experiences. Observation of the children is essential in order to make appropriate provision for individual learning. The work of Bruce (1987) and Nutbrown (1994) focus on and provide useful insights into schemas or patterns in children's learning and development. For example, children who pay particular attention to moving themselves or objects back and forth or up and down are likely to be interested in the levers and pulleys mentioned in Chapter 4. Those who like to dress up or cover things over may well enjoy cooking mince pies with lids on, or covering a surface when finger painting (see Chapter 3).

Observations do not need to be long or complicated, they can be written down on 'sticky' notes or any paper at hand, rather than the more formal record sheets previously referred to. These observations need to be more than a simple description of what the child is doing; they must provide information for the child's development (see examples below). All of the observations should be kept in a folder containing examples of a child's all-round development, and include drawings and possibly photographs, which are a very useful addition when working with this age range. Summative evaluation can then be carried out, maybe once a term, based on the observations gathered. The Desirable Learning Outcomes encourage teachers to make such observations of the children, by stressing the importance of collecting evidence from a range of activities over time.

Set out below are some examples of observations:

- One may make an observation of a child making a fan, which would be an example of folding for a purpose.
- Another child might cut out the animals from a piece of wrapping paper or an old birthday card, or be learning to use the hammer or saw at the woodwork bench, or building with wooden blocks – all examples of selecting materials and equipment for a variety of purposes.

Noting such things and collecting the observations builds up a picture of what each child can do within the context of spontaneous play.

- Jake at 3 years 10 months, about to hit the plastic bottles hanging on the washing line was heard to say;

  'move out of the way Daniel, I'm going to hit it very hard'.

Among other information, this indicated that he had an understanding of forces and had made an estimate of the length of the string, as well as how far and hard the bottles might travel. Very detailed observations are time consuming and not necessary. But knowing from earlier observations that Jake was particularly interested in things that moved forwards and backwards helped staff to make a range of provision to accommodate his all-round development, including science.

Observation and recording thus provide the information needed to chart a child's development and progress as well as feeding into short-term planning. Long-term aims will be covered by the overall scheme of work which will include Desirable Outcomes. Seen in this light the possibly demanding chore of observation and assessment is a very effective and enlightening way of informing practice.

Profiles

A profile should provide an overall picture of a child's development including social and emotional development as well as academic development. A detailed profile gives an emerging picture of individual children as learners, and this can then be used as a good basis for the reports teachers write for parents and colleagues. Profiles kept on individual children can also assist the record-keeping process, as they can help teachers with their planning, teaching and assessment by:

- focusing on individual children and their learning in terms of knowledge, skills, attitudes and experiences;
- showing individual children's development over time;
- enabling teachers to reflect upon each child's strengths, interests and learning needs.

It is necessary to make observations that inform you as to where children are in their learning and how they approach tasks. Observations that note children's work habits and current levels of achievement are important, e.g.:

- participant observation where you interact with children and reflect on their responses;
- detailed analysis of children's work.

It is important when gaining an overview of the child that more than their academic ability is considered. Therefore observations should be made of the child in a variety of contexts such as the playground, assemblies, wildlife area, classroom area and other contexts within the school. All of these will enable the teacher to have a good understanding of the whole child.

SCIENCE IN THE WHOLE CURRICULUM

Science will occur informally through children's work and aspects of play, e.g. when young children are involved in cooking or washing in the home corner, or when older children in history lessons consider the types of materials and structures used for buildings in Tudor times. It will be up to the individual teacher to decide whether to follow up the aspect of science when it occurs or whether to leave it for a later occasion. What is important is that the teacher does follow it up at some point as this values the ideas and issues the children have raised.

When planning the curriculum, many schools will look for natural links to enable common or complementary knowledge, understanding and skills to reinforce each other, e.g.:

- developing writing skills through work in science; providing the children the opportunity to write in a different genre;
- work on the water cycle in science linked with work on weather and rivers in geography;
- work on the appropriate uses of materials in design technology;
- the development of materials over time in history;
- the accurate use of measurement for a purpose in mathematics;
- investigating sounds and pitch in music;
- representing scientific findings through drawing or modelling in art.

Skills acquired in one subject can be applied or consolidated in the context of others. Work in one subject can provide a useful stimulus for work in other aspects of the curriculum. It is, however, important to keep work focused by limiting the number of subjects to be linked and to avoid contrived or artificial links.

Classroom display and atmosphere

Displays are a regular part of the primary classroom; they are there to provide an exciting and stimulating environment for the children to work in. Children take pride in and enjoy viewing aspects of their work displayed in their classroom and around the school. This enhances children's self-esteem by giving them a sense of worth, which in turn encourages the children to take a keener interest in the activities they are engaged in. A dynamic classroom which is full of life, displaying the ongoing products of children's work, develops a positive working atmosphere. However, displays are related to many aspects of the primary curriculum, particularly art, design work, history and geography topics; and science is not readily thought of as a subject for display. Yet science is in fact an excellent subject for display, particularly for interactive display.

Interactive classroom displays give children the chance to use odd moments for observing and investigating, and can also provide fruitful starting points for activities without necessarily taking up valuable class time. It is also valuable to use science displays for ongoing work in a problem-solving context. If, for example, the class is trying to discover which kitchen towel is the best 'mopper upper', you may set up the display and say this is what we have found out so far when water is spilt, but we are now going to carry out further investigations with kitchen oil, tea and coffee. These results could then be added when the investigations are completed: Further examples for display are shells and pebbles displayed not only in a dry setting, but also in water so that their colours show more clearly and possibly differently. Snail shells of various sizes but of the same type provide a chance for children to enquire through observation how a shell grows. Children can also get some idea of mechanisms from observing objects which can be taken apart. Occasionally slight problems can occur with interested and enquiring children. Investigating an interactive display on floating and sinking as part of a larger topic on water, one group of children siphoned out the water from the tank all over the classroom! Looking on the positive side, at least this meant they had learnt how to siphon, even if it was not quite what was intended.

And finally . . .

We hope these ideas will help you to be enthusiastic and enjoy the science you have planned, while encouraging the children to express their own ideas and views. Keep clear records of the children's development and ensure through your planning that you present the children with exciting and stimulating activities that give them a good 'hands-on' experience of science.

Appendices

1 BIBLIOGRAPHY

Association for Science Education (1990) *Be Safe*. Hatfield: Association for Science Education.

Association for Science Education (1996) *How to write a Scheme of Work*. Hatfield: Association for Science Education.

Athey, C. (1990) *Extending Thought in Young Children*. London: Paul Chapman.

Bartholomew, L. and Bruce, T. (1993) *A Guide to Record Keeping in Early Childhood Settings*. London: Hodder & Stoughton.

BBC Education (1994) *Simple Minds?* (Video).

Bruce, T. (1987) *Early Childhood Education*. London: Hodder & Stoughton.

Department for Education (1995) *Science in the National Curriculum*. London: HMSO.

Donaldson, M. (1992) *Human Minds*. London: Penguin.

Driver, R. *et al.* (1994) *Making Sense of Secondary Science – Research into Children's Ideas*. London: Routledge.

Gilbert, J. and Qualter, A. (1996) 'Using questioning and discussion to develp children's ideas', *Primary Science Review* **43**, 6–8.

Gould, S. (1984) 'Sex, drugs, disasters and the extinction of dinosaurs', *Discovery* March, 67–72.

Lundie, A. (1994) 'SATisfactory?', *Primary Science Review* **32**, 24–6.

Nutbrown, C. (1994) *Threads of Thinking*. London: Paul Chapman.

Office for Standards in Education (OFSTED) (1994) *Science and Mathematics in Schools: A Review*. London: HMSO.

Office for Standards in Education (OFSTED) (1996) *Subjects and Standards: Issues Arising for School Development Arising from OFSTED Inspections Findings 1994–5, Key Stages 1 and 2*. London: HMSO.

Osborne, J., Wadsworth, P., Black, P. (1992) *Primary SPACE Research Report: Materials*. Liverpool: University of Liverpool Press.

Osborne, J. *et al.* (1990) *Primary SPACE Research Report: Light*. Liverpool: University of Liverpool Press.

Osborne, J. *et al.* (1994) *SPACE Research Report: The Earth in Space*. Liverpool: University of Liverpool Press.

Russell, T. and Watt, D. (1992) *Primary SPACE Research Report: Growth*. Liverpool: University of Liverpool Press.

School Curriculum and Assessment Authority (1995) *Planning the Curriculum at Key Stages 1 and 2*. London: SCAA.

School Curriculum and Assessment Authority (1997) *Looking at Children's Learning*. London: SCAA.

Watt, D. and Russell, T. (1992) *Primary SPACE Research Report: Sound*. Liverpool: University of Liverpool Press.

Watts M., Barber, B., Alsop, S. (1997) 'Children's questions', *Primary Science Review* **49**, 6–8.

2 RESOURCES

Classroom use

Published schemes

Ginn Science (1990)
This is a comprehensive scheme which covers the age range of the primary school from reception through to Year 6. There is a special pack for reception, and then each year group

has a file containing background science for teachers, ideas for activities and materials which can be photocopied to use with children.

Bath 5–16 (1995)
This scheme provides a progression and coherence across Key Stages 1, 2, 3 and 4. There are files which provide detailed background information for teachers, with corresponding work cards for the children to use. The scheme is divided into three main areas: life, materials and physical processes, which cover all aspects of the National Curriculum.

Nuffield Primary Science (1995)
This scheme is based on the work of the research project Science Processes and Concept Exploration. There are 11 teachers' guides for each key stage which cover ideas and ways of developing children's ideas in science. The Key Stage 2 books also provide background science for teachers, and there is also a separate book to support teachers.

Ginn Star Science (1996)
This is a topic-based scheme which covers all aspects of the National Curriculum. There is also a very comprehensive guide for teachers addressing issues of planning, recording and assessment.

Science Connections, Longman (1997)
This newly published scheme provides background information for teachers, and a structured science programme for children using differentiated activity cards. It also addresses planning, recording and assessment.

Topic resources

Electricity
Understanding Electricity (UE), the Electricity Boards' education service, and the Institution of Electrical Engineers' (IEE) education service, each have a catalogue of resources for pupils of all ages, some of which are free, e.g. two electrical safety videos, *Flashback* and *Power House* (suitable for Key Stage 2), with an activity folder *First for Safety*, from UE, and a poster on circuit symbols from IEE.

Earth and space
The teacher support pack *Earth and Beyond* (ASE/Association for Astronomy Education 1997) has extensive resources listings including Planetaria and other places to visit, workpacks, suppliers of slides and prints, internet sites and recommended pupils' books: the following are a selection:

- Pupil booklets from the teaching schemes *Star Science* (Ginn), *Nuffield Primary Science* (Collins) and *Bath Science* (Thomas Nelson);
- *Starting Point Science* series (Usborne);
- *Spacewatch* series (Eagle Books);
- *First Starts* series (Watts);
- *Investigating Space* series (A&C Black);
- *I Wonder Why* series (Kingfisher);
- *The Young Astronomer* (Usborne);
- *How the Universe Works*, Heather Couper and Nigel Henbest (Dorling Kindersley);
- *Our Universe – A guide to What's Out There*, Russell Stannard (Kingfisher).

Stories for floating, sinking and boats
Key Stage 1
- *When Piglet is Surrounded by Water*, A A Milne;
- *The Cow Who Fell in the Canal*, P Krasilovski.

Key Stages 1 and 2
- *Who Sank the Boat*, Pamela Allen;

- *Mr Archimedes' Bath*, Pamela Allen.

Key Stage 2

- *Space Baby*, Henrietta Brandford.

Environment

Clue Books series of books on identification of plants and animals using keys (Oxford University Press).

Further study

Books

- Bartholomew, L. (1993) *Getting to Know You.* London: Hodder & Stoughton.
- Qualter, A. (1996) *Differentiated Primary Science.* Buckingham: Open University Press.

Videos

- *Making Sense of Science* – ten 30-minute topics including *Sound, Light, Forces, Electricity* from SPE – for practical classroom ideas;
- *BBC Primary Science – Teaching Today* – six 30-minute topics including *Forces and Energy, Electricity* from BBC Education – for developing concepts.

Kits

- *Teaching Today* – six professional development kits including *Forces, Electricity* from BBC Education – for hands-on learning about science concepts;
- Resources Pack and Video – *Circuit Training* from IEE – for basic circuit practical work.

Workbooks

- Three self-assessment guides: *Energy, Electricity, Forces* (National Curriculum Council).

3 EQUIPMENT

Essential equipment for Key Stage 1

Collections of as wide a variety as possible of natural and made materials, e.g. shells, feathers, rocks, soil, bones, woods, plastics, metals, papers, fabrics

Collection of wheeled and other moving toys for investigating force and movement

Ramp and stand

Electricity – not a commercial kit, but sufficient batteries, bulbs, wire, etc. to support investigative work in electricity

Musical instruments and materials to make a range of sounds

Range of magnifiers, e.g. hand lenses, binocular microscope, Fresnel lenses

Thermometers

Timers, e.g. sand timers, tockers, water clocks

Mirrors, e.g. plastic, flat and curved

Magnets

Globe to represent Earth

Plasticine

Assorted containers – for collecting and housing living specimens (plant and animal) together with paint brushes, spoons, tea strainers, etc. for handling, pots for growing seeds, large transparent water trays

Heating apparatus: large roasting tray, sand, lump of Plasticine, aluminium foil, candle, wooden dowel with peg attached (see Figure 3.6)

Wooden lolly sticks or tongue depressors

Sieve – a tea sieve would be appropriate

Cartesian diver (see Figure 3.3)

Desirable equipment for Key Stage 1

Electronic scales
Electronic timers, stop-clocks
Epsom salts
Measuring cylinders
Silly putty, jelly, clay, chocolate, talc
Force meters
Plastic guttering
Extra components to allow more complex electrical circuits, e.g. buzzers, switches
Wider range of types of thermometers, e.g. clinical, 'thermostik' type
Wider range of equipment for ecological work, e.g. dipping nets, enamel dishes, trowels, funnels, collecting pots with lids

Essential equipment for Key Stage 2

Electricity kits to include a range of batteries, bulbs, wire, screwdrivers, wire-strippers, buzzers, bells, switches, motors, clips for connectors (crocodile or Fahnstock type), large iron nail for making electromagnet
Heating apparatus: large roasting tray, sand, lump of Plasticine, aluminium foil, candle, wooden dowel with peg attached (see Figure 3.6)
Collections of as wide a variety as possible of natural and made materials, e.g. seeds, leaves, shells, feathers, rocks, fossils, soil, bones, woods, plastics, metals, papers, fabrics
Collection of musical instruments including some unusual ones
Collections of gears, springs, mechanical toys, balls
Magnets – variety of types, e.g. horseshoe, bar, rod, ring button, ceramic
Plotting and directional compasses; iron filings
Range of magnifiers including container magnifiers with measuring grid on the base, binocular microscope
Thermometers including indoor and outdoor types
Concave and convex mirrors and lenses, coloured filters (gels)
Timers including stopwatches, stop-clocks, digital timers
Equipment for ecological work, e.g. nets, measuring tapes, universal indicators/soil-testing kit, water-testing kit, pipettes, disposable gloves
Pots for growing seeds
Selection of household chemicals, e.g. salt, sugars, flours, bicarbonate of soda, vinegar, lemon juice, red cabbage water
Spring balances – a variety which measure in grams and newtons
Sets of slotted masses of 10 and 100 g
Materials for heat and sound insulation – foam and fabrics
Earth model mounted on inclined axis

Desirable equipment for Key Stage 2

Electronic scales, bathroom scales
Transparent plastic aquaria, guttering
Models of human ear, eye, torso, skeleton, etc.
Tuning forks
Ray boxes for producing light rays on tables, with cylindrical mirrors and lenses
Equilateral prisms, pinhole camera and kaleidoscope kits, optical fibre
Ramps and stands – various lengths and heights
Orrery of solar system.

Glossary

Absorption Uptake of substances into cells or tissues as in the absorption of digested food material
Acceleration The rate of change of speed or velocity: measured in metres per second per second
Adaptation A feature which makes an organism better suited to its environment
Amplitude The size of a vibration which produces a wave
Atom The smallest particles which are the building blocks of materials (From the Greek meaning 'cannot be cut')
Autotrophic nutrition Self feeding where carbon dioxide is the main source of carbon for building up fats, proteins and carbohydrates
Big Bang Theory of the origin of the universe, the start of an expansion which is still continuing
Black hole The result of the collapse of a large star; gravity is so large not even light can escape
Boiling point The temperature at which a liquid turns to a gas
Brittleness A material which breaks suddenly and completely is brittle, e.g. china plates and plain window glass
Burning A chemical change which requires three factors: a fuel, oxygen and a high temperature
Cell (electrical) This produces electricity by chemical action
Cellulose A carbohydrate found in plants where it acts as structural material; is found in all plant cell walls
Circuit An arrangement of conductors which allows a current to flow from an electricity supply
Compound A compound is made when two or more substances are combined to form a chemically different substance from the original ones
Compression Equal forces pushing on an object resulting in its being squashed
Conductor A material which allows the flow of electric current or heat
Current (electric) The flow of electrons (charge); measured in amp(ere)s
Discharge The loss of electric charge, often with a spark
Earth's axis The imaginary line through Earth from North to South about which the Earth rotates. This axis is inclined to the direction of Earth's orbit, and this causes seasonal variations
Elastic Able to change back to its original shape when the deforming force is removed, e.g. rubber ball
Electric charge The property of electrons
Electromagnet A coil of wire with an iron core which is magnetic when a current flows
Electromagnetic spectrum A family of waves which carry energy at very high speed, including light, X-rays and radio waves

Electron A very small negatively-charged particle which is part of the atom; responsible for electrostatic and current electricity effects
Electrostatics The effects of electric charge
Element Materials which consist of all the same atoms, e.g. carbon, bonded together to make the smallest part of compounds
Embryo Young organism in early stages of development
Energy The ability to produce a change: measured in joules
Frequency The rate of a vibration which produces a wave
Fruit In biological terms, a fruit is not necessarily something to eat – it is the fertilised ovary of a flower, e.g. in the pea or bean, the fruit is the pod and the peas or beans are the seeds within it; the ovule on fertilisation becomes a seed and as the seeds mature, the ovary increases in size to form a fruit
Galaxy A very large collection of stars; the solar system belongs to the Milky Way galaxy
Gas A gas has no shape of its own, but will fill the container which it is in; as the atoms or molecules from which it is made have no bonds (forces) between them, they are free to move in all directions
Generator (electrical) An electromagnetic device which makes electricity from movement
Gravity The force of attraction between all objects that have mass; results in objects on Earth falling towards the centre of Earth
Heterotrophic nutrition Feeding on others; here, organic substances – fats, proteins, and carbohydrates – are derived from taking in other organisms which are digested and absorbed
Insulator A material which prevents the flow of electric current or heat
Internal energy The energy inside an object which affects its temperature and its state
Kinetic energy The energy that a moving object has; the more massive the object or the faster it is moving the more kinetic energy it has
Liquid A liquid has no shape of its own and will take up the shape of the container that it is in; the atoms or molecules from which it is made are not strongly bonded and are able to move over and around one another allowing the material to 'flow'
Melting point The temperature at which a solid turns to a liquid
Mixture A mixture is made when two or more substances are physically mixed, while the original substances remain chemically discrete
Molecule A group of atoms, e.g. molecules of water would have three atoms: two hydrogen and one oxygen
Multimeter An instrument for measuring electrical quantities; current in amp(ere)s, voltage in volts and resistance in ohms
Ohm's law When a metal component is in a circuit at a constant temperature, the current that flows through it is proportional to the voltage across it; this is called its resistance
Organ A structure such as the heart, brain or kidney, which consists of a number of tissues acting together to perform a complex major function

Organism An independent living unit; it may be a single cell, as in an amoeba, or made up of many, as in ourselves

Ovary Female reproductive organ of flowering plants; it contains one or more ovules and, after fertilisation, the ovary wall may form the outer coating of a fruit, e.g. pea pod

Ovule On fertilisation this becomes the seed and the whole ovary will form the fruit

Parallel circuit A circuit in which the components are arranged to provide alternative routes

Parasitic nutrition A type of heterotrophic nutrition in which one organism derives food from the living body of another organism of a different species

Photon A particle of electromagnetic wave energy; ultraviolet photons have more energy than light photons which is why they can damage the skin

Plastic (behaviour) Unable to change back to its original shape when the deforming force is removed, e.g. Plasticine

Plastic (substance) A material with organic structure, originating from oil or similar substance; note that many plastic substances behave elastically

Pollination The transfer of pollen from the male reproductive organ (the stamen) to the receptive area (stigma) of the ovary; a plant may be cross-pollinating or self-pollinating: cross-pollinating is where the pollen which fertilises the ovule is from another plant of the same species; self-pollinating is where the pollen which fertilises the ovule is from the same plant

Potential energy The energy of position, e.g. something high up has potential energy because of gravity

Power The rate of energy transfer: measured in watts; a watt is a joule per second

Pressure The force on a unit area of a surface: measured in newtons per square metre

Radiation energy The energy of electromagnetic waves which includes 'heat' (infrared), light and microwaves

Reflection The process of bouncing back off a surface at a predictable angle, e.g. light reflects regularly off a mirror

Refraction The process of changing direction by travelling through a different medium, e.g. light refracts in glass or water. This is caused by a change in the speed, and can produce images and spectra

Resistance Opposition to flow, e.g. electrical resistance, air resistance

Rusting A chemical reaction between iron, oxygen and water which results in the formation of iron oxide and water, with the effect of corroding or 'eating away' the metal

Saprophytic nutrition A type of heterotrophic nutrition in which an organism is able to digest organic material outside its body through secreting enzymes; the digested material is absorbed back into the body

Satellite An object which orbits another, e.g. TV broadcasting satellites orbit Earth

Series circuit A circuit in which all the components are arranged one after another

Solid An object is a solid which retains its own shape; the atoms or molecules from which it is constructed are bonded closely and strongly together and, although vibrating, cannot move apart from one another

Solute The substance that dissolves, e.g. sugar in water

Solution A solution is clear; if left to stand, it will never settle out and will not separate when filtered

Solvent The liquid that something dissolves in, e.g. the water in brine

Speed The distance travelled in a unit of time: measured in metres per second

Stamen The male reproductive organ of a flowering plant. It consists of a delicate stalk called the filament which has an anther at the end. The anther produces pollen and, on ripening, bursts open to release the pollen prior to pollination

Star A very large object which radiates energy from the nuclear reactions which power it

Strength A measure of the force which an object can withstand before breaking

Suspension A suspension is cloudy; if left to stand it will settle out to give a sediment and can be separated by filtering

Temperature A measure of the 'hotness' of an object, related to the internal energy it contains: measured in degrees celsius (°C) with a thermometer

Tension Equal forces pulling on an object resulting in its being stretched

Transformer An electromagnetic device which is used to change the voltage of an electric current

Upthrust The force on an object in a fluid (liquid or gas) which helps to support it

Velocity The distance travelled in a straight line in a unit of time: measured in metres per second

Viscosity The stickiness of a material, usually a liquid, caused by the resistance of the molecules to move; treacle is more viscous than water at room temperature.

Voltage The potential to make an electric current flow, e.g. a cell has a voltage of 1.5 v

Wavelength The length of a wave from one crest to another

Index